POLYMER SCIENCE AND TECHNOLOGY SERIES

THERMOSETTING PLASTICS – 2nd edition
MOULDING MATERIALS AND PROCESSES

POLYMER SCIENCE AND TECHNOLOGY SERIES

SERIES EDITORS: DR D M BREWIS AND PROFESSOR D BRIGGS

Published

H BRODY (ed.), *Synthetic Fibre Materials*

E A COLBOURN (ed.), *Computer Simulation of Polymers*

F R JONES (ed.), *Handbook of Polymer-fibre Composites*

I S MILES AND S ROSTAMI (ed.), *Multicomponent Polymer Systems*

D E PACKHAM (ed.), *Handbook of Adhesion*

ROGER ROTHON (ed.), *Particulate-filled Polymer Composites*

Forthcoming

D M BREWIS, B C COPE AND R DYSON (eds), *Handbook of Polymer Science*

POLYMER SCIENCE & TECHNOLOGY SERIES

THERMOSETTING PLASTICS – 2ND EDITION

MOULDING MATERIALS AND PROCESSES

J F MONK
with special contributions by
P D COATES, A F JOHNSON and W A LEES

LONGMAN

Addison Wesley Longman
Addison Wesley Longman Limited,
Edinburgh Gate, Harlow,
Essex CM20 2JE, England
and associated companies throughout the world

First edition published 1981 by George Godwin Ltd
the publishing subsidiary of The Builder Group
in association with
The Plastics and Rubber Institute

This edition published 1997

British Library Cataloguing in Publication Data
A catalogue entry for this title is available from the British Library

ISBN 0-582-04071-X

Library of Congress Cataloging-in-Publication Data
A catalog entry for this title is available from the Library of Congress

Set by 4 in 10/12½ Times
Printed in Great Britain by Henry Ling Ltd., at the Dorset Press,
Dorchester, Dorset

Contents

Preface

The obvious question facing every potential author is – why write a book?

When the first edition of *Thermosetting Plastics* was published in 1981, it was at the end of a period in which thermosetting moulding materials had successfully initiated a transition from processing by the traditional compression and transfer moulding to now include the injection moulding process, which had already come to dominate the rest of the plastics moulding industry.

As is witnessed in every other field of our modern world, much has changed during the intervening years. The plastics industry, like most others, has had to re-invent itself to maintain and improve its position in an increasingly competitive world.

The changes that have occurred are not only in the essential materials technology and advancement of the moulding technologies. Because the whole manufacturing environment is changing rapidly, new issues have arisen that challenge manufacturing industry to become more efficient and responsible in every way — not only in terms of higher output and low costs, but also in issues such as efficient use of precious resources, environmental soundness, life cycle useage of materials and numerous others.

At the same time, applications for plastics materials have continued to become more demanding in terms of product performance, working environments, product life and overall quality standards.

In this highly competitive environment, we can clearly see that thermosetting moulding materials have continued to gain ground in overall terms, and are being increasingly turned to in application areas that combine the needs to have product performance in demanding environments and at very competitive price levels. With these attributes it is not surprising that, for example, thermosetting materials are extensively used in automotive applications, and many advances in materials performance and fabrication processes have been driven by the needs of this industry.

The purpose and justification for this book has therefore become increasingly clear.

There is still a need to provide an up-to-date guide and reference work on thermosetting moulding materials and processing technologies. It is also essential to address, in one volume, the fundamental characteristics, performances and behaviour of the main thermosetting families and then provide a systematic and detailed exposition of the main production processes.

The book is organized to meet these objectives, with the treatment of thermosetting materials in the first part, including the essential polymerization mechanisms and rheological background to enable a full understanding of their behaviour during processing. This part of the book is completed by a series of tables which summarize the material properties and processing conditions for each type of compound.

The second part of the book is then devoted to a detailed description of the main fabrication processes, processing equipment and moulds, with practical advice on the setting up, fault finding, optimization and control of the processes.

Special chapters are devoted to the processing of polyester bulk and sheet moulding compounds, and on the reaction moulding processes that are now becoming very important sectors of activity.

Finally, a chapter has been included which deals with the important, and often overlooked, aspects of finishing, treatment and assembly of thermosetting mouldings.

A wide range of accumulated experience and expertise has been drawn on in putting this work together. In particular, I have been fortunate in having the chapter on reaction moulding technology provided by Professors Coates and Johnson of the University of Bradford, who have played a significant role in the evolution of this field. Dr Lees has also kindly provided the section on adhesive bonding in Chapter 9.

It is to be hoped that this book will be of value to the practising production and processing engineer, materials technologists and students of polymer engineering and technology.

In hoping that this work will be welcomed as a useful addition to the literature in this subject area, I should also like to record my thanks to the many friends and colleagues in the industry who have contributed both information and advice, as well as the editorial staff of Addison Wesley Longman for their patient and professional assistance in producing the final product.

J F MONK

Acknowledgements

We are grateful to the following for permission to reproduce copyright material:

Carl Hanser Verlag for figures 6.3 and 6.4 from **Johannaber** (1993), *Injection Moulding Machines (3e)*, and for figures 7.20 and 7.21 from **Kia** (1994), *Sheet Moulding Compounds*; PPG Fiber Glass for figure 4.6 from **J. Klunder** (1993), *PPG Fibre Glass technical manual on glass fibres (2e)*; The British Plastics Federation for figure 1.25 from *Conference Paper No. 20: BPF Composites Congress 1992*.

We acknowledge the contribution made by Dr W. A. Lees of the Permabond Division of the National Starch & Chemical Ltd to Chapter 9.

We also thank those individuals and organisations who supplied illustrative material for use in this book.

While every effort has been made to trace the owners of copyright material, in a few cases this has proved impossible and we take this opportunity to offer our apologies to any copyright holders whose rights we may have unwittingly infringed.

CHAPTER 1

Fundamental considerations

1.1 Introduction

As a result of the substantial improvements in the range and performance of thermosetting moulding materials available today, plus the dramatic advances that have been made in moulding processes, thermosets are increasingly finding new applications. The growing need for polymer products in demanding environments and the requirement to deliver high performance at competitive price has ensured that thermosetting materials continue to be specified and used in a vast array of moulded products.

This book has been compiled with the aim of bringing together in one volume a comprehensive compilation of materials and processing information over the broad spectrum of thermosetting moulding materials. It will help the reader specify, design and manufacture high-quality products benefiting from the accumulation of knowledge gained over many years of experience.

Emphasis is given to thermosetting materials and techniques currently available for volume production of thermosetting mouldings. The first part of the book deals with the basic characteristics, properties and applications of thermosetting moulding materials in general terms and also on a family-by-family basis. The second part of the book deals with the technology and application of the main moulding processes. Where such processes are specific to particular thermosetting materials families, for example thermosetting polyesters and reaction moulded polyurethanes, the specific chapters include a more in-depth treatment of the performance of the materials in this particular context.

The designer, materials engineer and manufacturing engineer now have available a multitude of polymer resins, compounds and systems from which to select the most appropriate option. The choice can often be bewildering and daunting. While there are considerable numbers of books and much detailed trade literature dealing with the use of thermoplastic polymers, pertinent information relating to thermosetting plastics is much harder to come by. As a result it is not uncommon that thermosets are not considered at the selection stage. As this book sets out to show, this error can prove to be costly, restrictive and potentially uncompetitive in many cases. Thermosetting materials can offer a unique blend of properties and performance at a very competitive price. It is to be hoped that this book, in dealing with an area of plastics technology that has been somewhat neglected in the technical literature over the years, can provide the necessary background to support new applications and improvements in the use of these materials.

1.2 Basic characteristics of thermosetting resins

While this book does not explore the in-depth chemistry and physics involved in the manufacture and use of thermosetting moulding compounds, which is covered elsewhere,[1,2] it is essential that the reader is provided with an adequate explanation of the key characteristics of this family of materials.

1.2.1 Polymer structure

Thermosetting resins, like thermoplastics, are synthetic materials built up by organic polymer molecules of high molecular weight. They differ in the structure of their molecules, with thermoplastics having more or less linear molecules, whereas thermosets are formed of complex three-dimensional densely cross-linked network-like structures of very high molecular weight.

1.2.2 Cross-linking by chemical reaction

The formation of thermoset cross-linked three-dimensional networks is the result of an irreversible chemical reaction, often referred to as curing. It is usually initiated by heat, and pressure is commonly needed to consolidate the cured polymer and to contain volatiles released during the curing reaction. The resultant networks are infusible and rigid by virtue of the chemical bonding that has occurred, and will keep dimensionally stable at increased temperatures until a sufficiently high temperature is reached at which oxidative breakdown of the polymer commences.

This behaviour contrasts sharply with the ability of thermoplastics to soften and flow with increasing temperature, and then re-solidify when cooled. In this reversible process no chemical reaction occurs.

1.2.3 Curing shrinkage

A feature of the thermoset curing reaction is that dimensional changes may occur both during and after moulding. This shrinkage occurs because the formation of chemical bonds between atoms while curing allows them to approach closer to each other than when non-bonded. This accounts for part of the mould shrinkage. Other causes of shrinkage are differences in coefficient of thermal expansion in comparison with steel used for the moulds, and loss of volatiles during the curing operation. Some volatiles only escape slowly by diffusion after moulding, and molecular segments of the resin matrix release built-in tensions by contraction, all of which contribute to post-moulding shrinkage.

1.2.4 Compounding with fillers

Because the cured thermosetting resins often tend to be not only rigid and hard but also brittle, it is common practice to compound the resin system, in the partly reacted (pre-polymer) stage, with a variety of fillers and reinforcements to provide improved mechanical and other specific properties in the finished product after the cross-linking (curing) reaction has been completed during the moulding process.

This is a major aspect of thermosetting moulding compound technology and is dealt with in depth during subsequent chapters. The amounts of filler incorporated in the moulding compound can be very substantial, and often well in excess of the amount of resin within the formulation. This can mean that not only are the properties of the final compounds significantly modified, but the total cost of the moulding compound can be substantially different to the basic resin price. As a result, thermosetting compounds can be designed that are very cost effective, since many of the fillers are much cheaper than the resins employed. This can often prove a considerable advantage in competition with thermoplastics, where the moulding materials are often undiluted resin, or modified by the use of a single filler or reinforcement at correspondingly lower addition levels than thermosets.

1.2.5 Flow/cure behaviour

The moulding of thermosetting moulding compounds is complicated by the fact that two competing processes occur at the same time. On the one hand, application of heat causes a softening of solid resins and a viscosity reduction of the melt phase. In competition with this physical effect, the application of heat also activates the cross-linking reaction, which starts with thickening of the melt and results in rigid cured polymer. Hence we talk about the flow/cure behaviour of thermosets, and can illustrate the net effect of the two processes with a flow/cure graph of the type shown in Fig. 1.1.

There are a number of standard tests developed to measure and control the flow/cure behaviour of the compound, some based on the distance the compound will flow under standard conditions within a mould with a spiral channel. Others, such as the Brabender Plastograph, measure the torque required to rotate two specially designed rotors at constant shear rate within the curing material. These tests are discussed in greater detailed in Chapter 2.

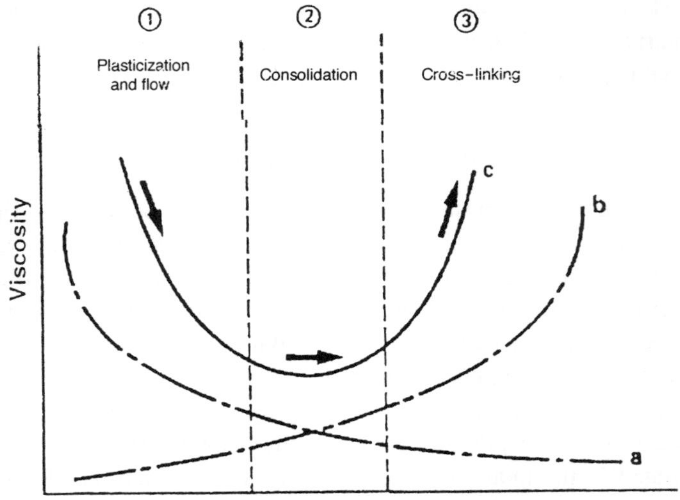

Figure 1.1 Generic flow/cure behaviour of a thermosetting moulding compound.

1.2.6 Key property areas

If the production of components from thermosetting moulding materials is not altogether simple in comparison to other materials such as thermoplastics, why is it that thermosets have always, and will continue to gain, favour in so many important application areas?

There are many powerful arguments to support their cause. First, and most important, as a result of their cross-linked infusible structure thermoset products possess unrivalled thermal resistance, stiffness, creep resistance, dimensional stability and chemical resistance in relation to other organic polymer families. In addition, they offer an excellent combination of electrical properties, being particularly resistant to high-voltage electrical tracking.

Most importantly, these properties are provided by families of materials which in general terms are relatively inexpensive when compared with thermoplastics offering the same thermal, mechanical, chemical and electrical performance. This benefit results from a number of factors. The resin systems themselves are relatively low cost, the moulding compounds, as previously mentioned, often contain large percentages of relatively cheap fillers and reinforcements, and finally thermosets compare very favourably with other materials with respect to the total energy requirements for the basic polymer feedstock and manufacture of the polymer. Figure 1.2 shows the energy requirements for manufacture of products from a range of common constructional materials. The energy requirements for feedstock (where appropriate) and fuel to manufacture the raw material are shown

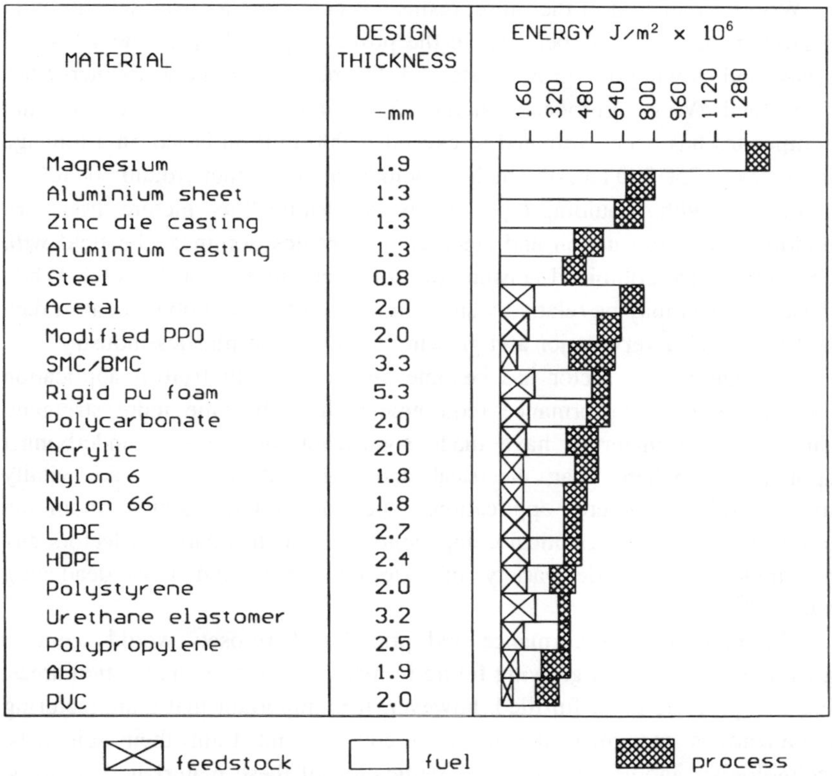

Figure 1.2 Energy requirements of various materials at typical design thickness.

separately from the process energy required to fabricate the product. Figure 1.2 also gives an indication of typical design thicknesses applicable, based on considerations of stiffness and processability.

1.3 Market summary and prospects

Thermosets have a long and distinguished history, and yet still promise an excellent future thanks to the constant advances in resin technology, fillers and reinforcements and the development of new and more effective moulding processes.

It was in 1909 that Leo Baekeland produced a resin from phenol and formaldehyde which, when mixed with fillers, could be cured under heat and pressure into a thermosetting plastic. This was the first completely synthetic plastic and became the basis for a plastics industry in general.

Today, the range of thermosetting moulding materials has widened enormously from the phenolics to include urea, melamine, melamine-phenolic, polyester, epoxy, alkyd, diallylphthalate, silicone, urethane and others.

World production of thermoset resins has continued to grow steadily with a growth rate of 8–9% per year in the post-war period up to the 1974 'oil crisis' and slowing to a growth rate of 4–5% per year in the years thereafter. The total West European market for fibre-reinforced thermosetting composites has been estimated to exceed £1000 million based on a tonnage in excess of 750 000 tap, of which it could be assumed that around 30–40% is associated with moulding-type processes. Within this picture transport, building and construction and electrical/electronics dominate the field with an approximate combined consumption of three-quarters of the total. While these numbers may be relatively small in relation to total polymer sales, they still represent a very major and growing sector of the plastics industry.[3]

The automotive sector has become an increasingly fruitful application area, as product performance requirements have become more stringent. Thermosetting materials have made significant inroads into underbonnet applications, where heat, electrical and chemical resistance are vitally important, in structural applications, such as bumper beams, mounting brackets and supports, optical components, such as headlamp reflectors and housings, and a wide variety of external panels and noise-deadening panels.[4,5]

The properties, performance and price of thermosets would seem to guarantee a bright and growing future in many demanding application areas. For this promise to be fulfilled, however, it is important that manufacturing organizations recognize the need to educate and train their scientists, technologists and engineers in the application of these materials. It is to be hoped that this book plays some small part in providing the necessary knowledge base.

1.4 Main conversion processes

It has already been stated that the purpose of this book is to focus on the conversion processes used for volume production of thermosetting mouldings. As a result, no attempt is made to cover the diverse techniques of hand lay-up, spray lay-up, continuous processes, including extrusion and pultrusion, filament winding and other specialist composite technologies. The main emphasis is placed on treatment of injection moulding, transfer moulding and compression moulding of granular, dough and sheet materials, as well as a separate chapter on the growing sector of reaction moulding processes.

The correct selection of the moulding process will depend on many factors. Product shape, size, property requirements, quantities, complexity, type of material to be processed and plant availability are some of the factors affecting the decision and hence the economics of manufacture.

Each major process available is covered in detail within the following

Table 1.1 Main advantages and disadvantages of the principal conversion processes

Advantages	Disadvantages
Compression moulding	
1. Can mould heavily reinforced materials	1. Labour intensive, with more material handling involved
2. Best method for mouldings with large surface area related to thickness	2. Material losses due to the need for flash allowance and general wastage
3. Greatest uniformity in moulded properties (i.e. isotropy)	3. Secondary operations for flash removal are generally required
4. Most versatile process with relation to quantities and component complexity	4. Poor control of part thickness and weight
5. Simple to understand and control	5. Difficult to mould round delicate inserts or pins
Transfer moulding	
1. Improved labour utilization compared with compression moulding	1. More skill in terms of process control is required
2. Shorter cures often achievable compared with compression moulding	2. Moulds often more complex and costly
3. Good control of part (and flash) thickness	3. Cannot use as wide a range of compounds as with compression moulding
4. Suitable for incorporation of delicate inserts	4. Greater risk of distortion and directionality of mechanical properties (i.e. anisotropy)
	5. Care often required to minimize effect of weld (knit) lines
	6. Material losses in central feed system may be incurred
Injection moulding	
1. Process offers scope for automatic operation	1. Greater skill required in setting up and controlling the process
2. Rapid cycle times result from a substantial reduction in cure time	2. Machines and moulds are more expensive and complex
3. Good control of dimensions and flash thickness	3. Process is not tolerant to stop–start conditions of operations
4. Good part-to-part consistency	4. More fibre orientation and degradation than compression moulding
5. Can mould thin sections and (preferably with vertical clamp units) around delicate inserts	5. Care is often required to minimize effect of weld (knit) lines
	6. Irrecoverable loss of materials in sprues and runners*

* Development of warm runner moulds to obviate these losses is covered in Chapter 6.

chapters. However, it is considered useful to list in Table 1.1 some of the main advantages and disadvantages linked to compression transfer and injection moulding.

The comparison between the relative characteristics, benefits and limitations of each conversion process provides an ever changing picture, as advances in materials, moulding processes and moulding process ancillaries occur.

For example, for some years there appeared to be a relative stagnation in the traditional compression and transfer moulding process technology, at a time when rapid evolution of injection moulding was occurring. Now we can see that many technical advances pioneered during the evolution of the injection moulding process (such as improvements in process setting and control, and better performing hydraulic components) are now finding their way into compression moulding technology. Much work is also being done to accelerate the process of compression moulding by introducing the pre-warming of the moulding compound.

An examination of the evolution of the injection and compression moulding materials and processes for the manufacture of large-area mouldings with unsaturated polyester compounds demonstrates this point very effectively.[6]

1.5 Product design guidelines

While many measures can be taken to predict and control the quality of components and finished products during and at the end of the various stages of manufacture, there can be no question that the design phase has by far the greatest influence on product quality, competitiveness, ease of manufacture, and all-round effectiveness of the manufacturing operation.

This may be easy to comprehend as a general principle, however, it is not always easy to achieve high standards of effectiveness in product design in practice. It can be suggested that the product designer has three main objectives to achieve from his work:

1. to design a product that is 'fit for purpose';
2. to design making effective use of the selected materials;
3. to ensure that the designed product is suitable for manufacture by appropriate processes and/or manufacturing systems.

Designing in plastics to achieve these objectives can present particular difficulties, since there is often a shortage of the engineering property data on commercial plastics, as distinct from property data that can form the basis of a purchase and quality control specification. However, in many cases the problem is not a lack of knowledge or understanding of how plastics behave, since the plastics and polymer literature is now becoming quite extensive with regard to plastics behaviour under a variety of stresses. Rather, the problem is a lack of a system for adapting true performance properties of plastics to a general approach to design which has evolved to a large extent around the theory of elasticity. Allied to this is the problem that the majority of designers have to treat polymers as one of many (diverse) materials, and cannot be expected to carry out in-depth research on such fundamental topics.

In these circumstances it is essential that there is a close liaison between the designer, the sales engineer, the materials engineer, the manufacturing engineer and the costing engineer/estimator within an organization, plus contact with the end user and materials supplier as necessary. It is clear therefore that teamwork is of the essence during the design process.

1.5.1 Systematic materials selection

As stated above, one of the primary objectives during the design phase is to ensure efficient use of materials.

To be effective and consistent, design and materials selection must be based on more than just an identification of performance properties and routine methods of testing them. It must also provide a system for use in the process of developing applications, since plastics require many more different performance properties than the traditional materials they compete with, and also there is often a shortage of design equations and properties to substitute in them for the types of stress to be encountered by the product in service.

As a result, the materials selection procedure can be quite complex. One very systematic approach[7] recommends adoption of a sequence which involves defining all part requirements in application terms and then selecting valid materials performance properties that reflect these requirements. A typical sequence can then involve some or all of the following stages.

Stage 1 Conception of the part. Preliminary functional design. Selection of the basic fabrication process/route.

Stage 2 Screen candidate materials on basis of processability, available engineering properties that relate to the likely stresses which part's function will impose on the material, and cost.

Stage 3 Preliminary analytical design, that is calculation of wall thicknesses and other part dimensions using design properties where possible and experience where not.

Stage 4 Performance simulation, structural analysis, processing simulation and/or prototyping and functional testing under actual or simulated conditions of usage, as found necessary to optimize the design.

Stage 5 Finalization of design and materials selection on basis of analysis, testing and fabricated part cost.

Stage 6 Specification of the selected material.

Obviously some telescoping of these stages can be achieved when part requirements are simple or the part is similar to items previously manufactured. However, with the increasing application of computer-based design systems it is now becoming common practice to model part performance (mechanical, vibrational, etc.) using a Finite Element Analysis software package, and to also simulate the moulding of the part, enabling production of mould-filling conditions, and highlighting potential filling, shrinkage and distortion problems with other complementary software packages often capable of working directly from the same defined CAD mathematical and geometric model. This approach offers great potential to optimize the design process and lead to designs that are 'right first time', thus minimizing or eliminating the need for extensive prototyping. It should also be noted that rapid developments are also occurring in processes to enable rapid manufacture of prototypes, working directly from the core CAD part definition and eliminating the need to manufacture prototype metal moulds for the production of the prototype components.

There are now a number of computer software packages available to assist in the short-listing and selection of candidate plastics materials that are particularly appropriate for use at stage 2 of the process outlined above.[8,9] One such system, PLASCAMS, which is marketed by RAPRA Technology Limited, is very suited to this task since it contains a very comprehensive database covering all the main families of polymeric materials, which can be used to search for specific detailed materials property requirements, or alternatively to provide suggested candidate materials against a 'basket' of general performance characteristics that can be uniquely selected and ranked with respect to the needs of a specific application.

1.5.2 Product failure analysis and prevention

In conjunction with a more systematic approach to materials selection discussed above, attention is also focused on a growing trend to use systems and protocols at the design and product development stage whose aim is to identify potential modes of product failure and eradicate them before entering into production. The most well known of these systems is called Failure Mode and Effect Analysis (FMEA) and has become widely used in the automotive and aerospace industries.[10] In this procedure, the designer and the manufacturing engineer tabulate the potential modes of failure of the product (usually drawing on past experience as well as fundamental analysis) as a result of weaknesses in design and potential problems and failings during manufacture. Alongside each failure mode, the potential causes are recorded, and the potential frequency, severity and ease of detection of the failure are given a numerical rating. A ranking of potential failures is then carried out in relation to the potential failure risk (i.e. highest risk = highest potential frequency × greatest severity of effect × highest difficulty of detection, computed as a product of the three ranking factors).

The designer can then address the elimination of potential failure causes on a priority basis. This approach may seem cumbersome; however, the accumulated recorded analysis, information and results soon become an invaluable 'bible' for the product and its successors.

1.5.3 Specific design features

The final aspect of the process of achieving good and efficient designs is to ensure that the new design embodies the accumulated experience of the materials and processes to be used, in this case thermosetting plastics, and that the design has abided by the basic rules and guidelines that have developed over the years.

Since thermosetting plastics moulding has been practised for many years, there is much knowledge to be drawn upon. However, there are few companies that now have specialist experience in the area of technology, and as a result the key information is not always easy to come by.

The following notes and figures bring together a wealth of experience in this particular subject area, and the form of illustration has been chosen to clearly highlight for the reader the essential differences between 'good' design practice with thermosets and 'bad' design practice.

1.5.3.1 Design for simplicity and general ease of production

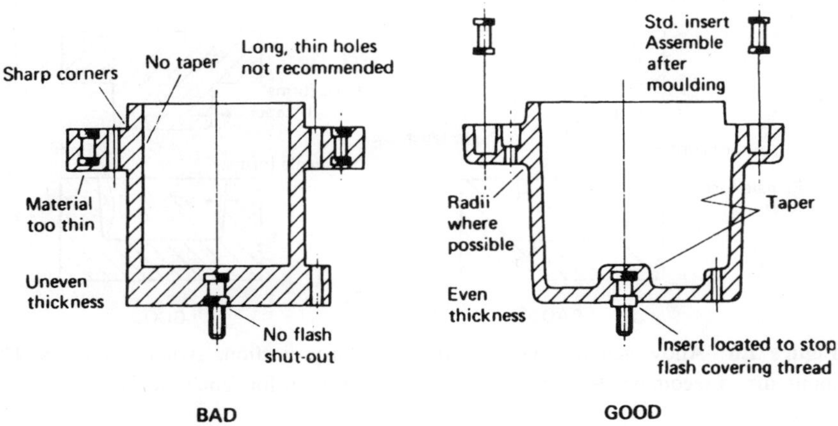

Figure 1.3 Simple mouldings are good mouldings and easier to produce. Bear in mind that irregular hollow shapes are unstable — they will distort during shrinkage as a result of the stresses that arise.

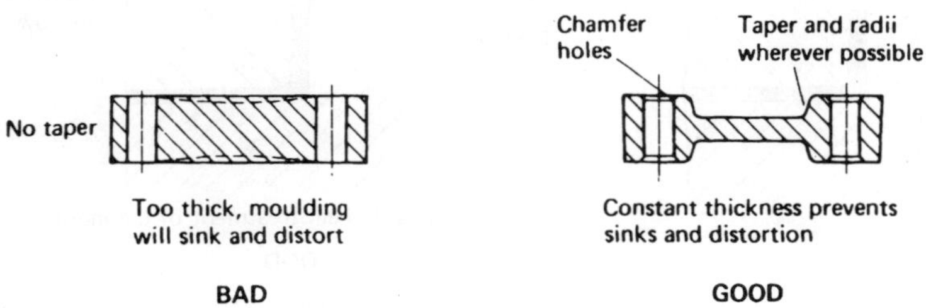

Figure 1.4 Section thicknesses must be kept as uniform as possible.

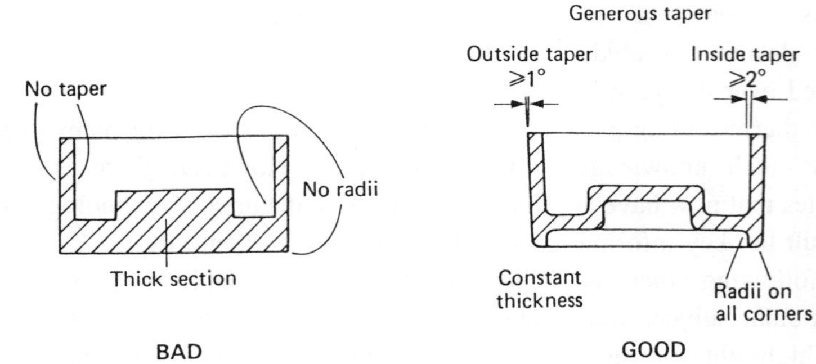

Figure 1.5 Avoid thin walls adjacent to thick sections: unequal sections induce stress and possibly cracking. Design radii as large as possible.

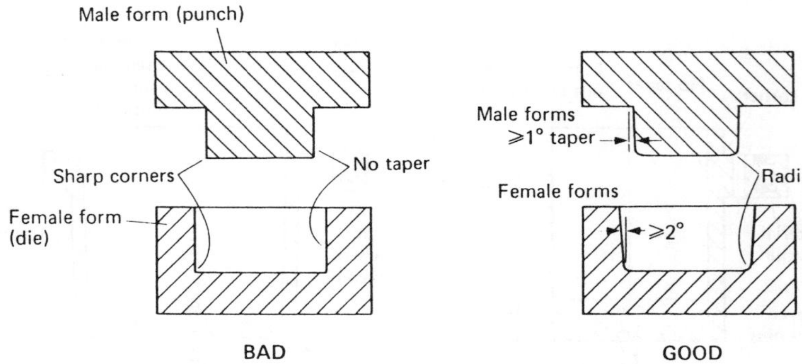

Figure 1.6 Allow generous taper or draw for easy ejection, avoiding stresses. 1° minimum is recommended for male forms. 2° minimum for female forms.

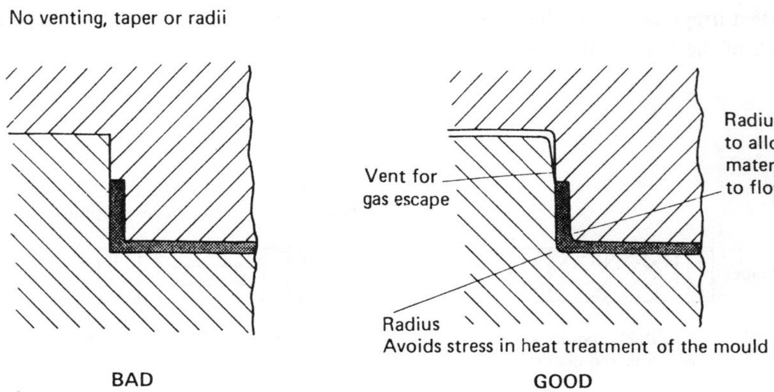

Figure 1.7 Largest possible radii in corners assist the flow of material and avoid stress in the female corners of the mould (which may have to be heat treated).

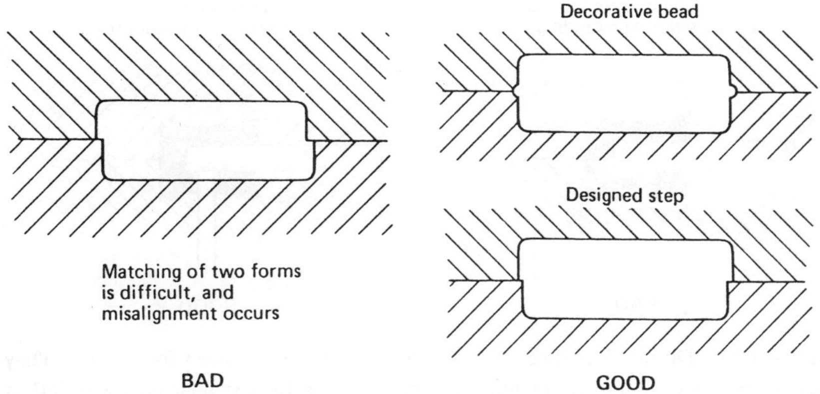

Figure 1.8 There is bound to be flash at the parting line of the dies, particularly when the tool wears. Arrange for easy removal if possible and/or where flash is not critical. Where there are matching forms in each half of the mould, a designed step or bead can be used to disguise any misalignment.

1.5.3.2 Incorporation of moulded holes, thread forms and undercuts

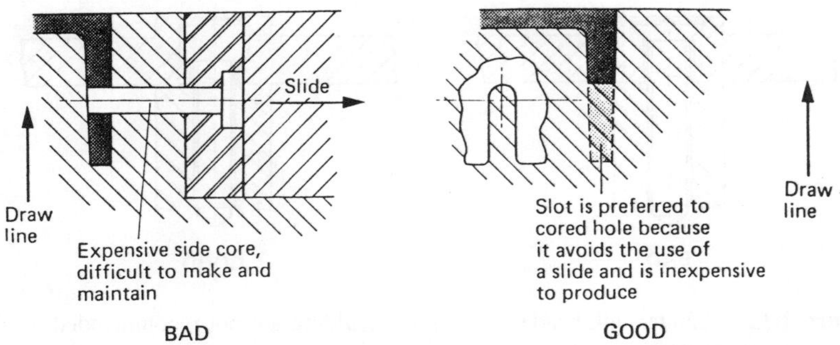

Figure 1.9 Dies normally move in opposite directions, perpendicular to the split line. Side holes, oblique holes and undercuts introduce complications. It may be cheaper to machine after moulding.

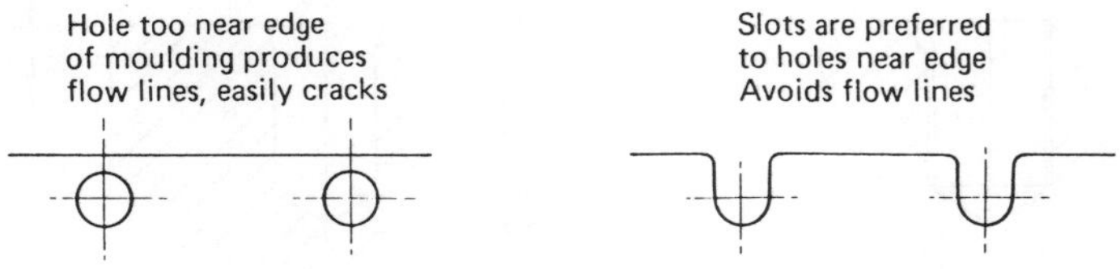

Figure 1.10 Do not locate holes near the edge or face — the thin sections adjacent to the hole will blister or crack. Can they be designed as slots to top or bottom of the moulding?

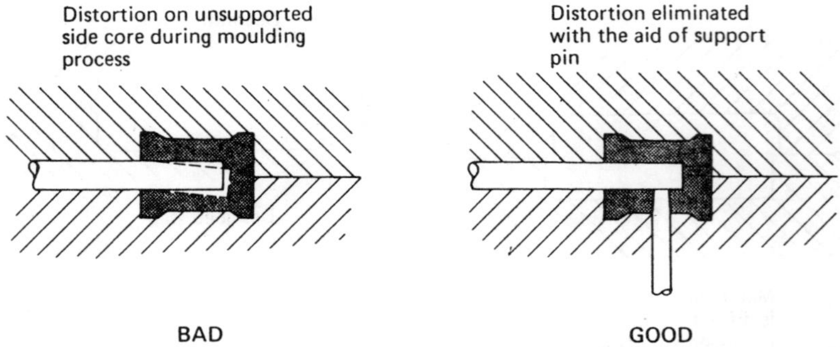

Figure 1.11 Do not introduce long side holes with no provision for support. They may be moulded by transfer or injection methods, but long slender pins will deflect under the pressure of injection. If the pins are too hard, they may break: if they are not hard enough, they will bend.

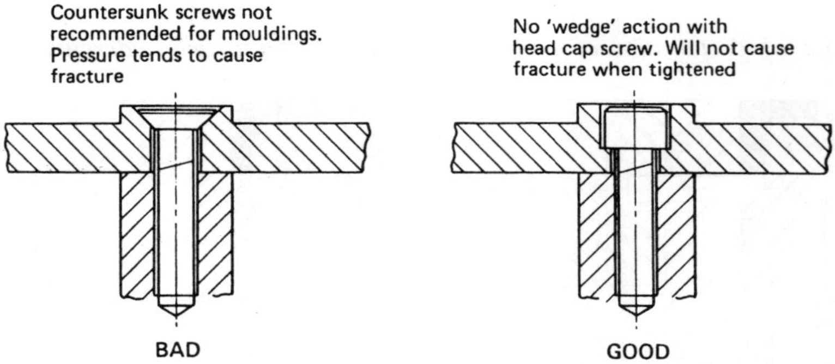

Figure 1.12 Countersunk headed screws in moulding are not recommended — the pressure will break the moulding.

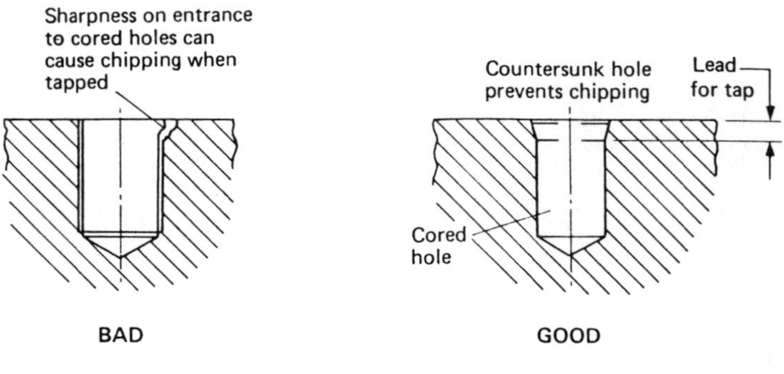

Figure 1.13 A countersunk entrance hole prevents chipping where a hole is to be tapped.

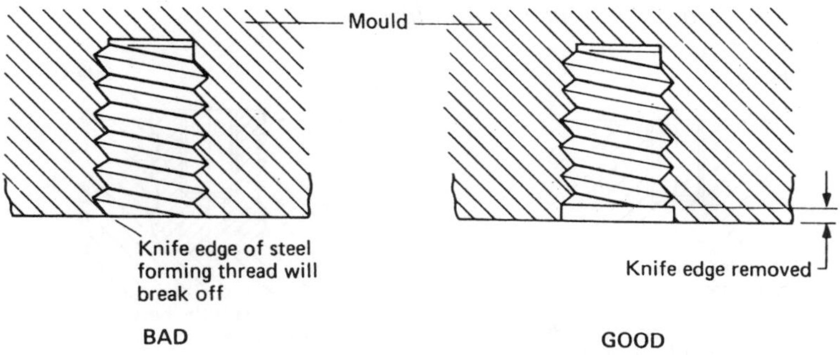

Figure 1.14 Do not design a male thread all the way down to the face of the moulding — it causes a feather edge on the steel where the thread runs out, and will be a continual source of trouble.

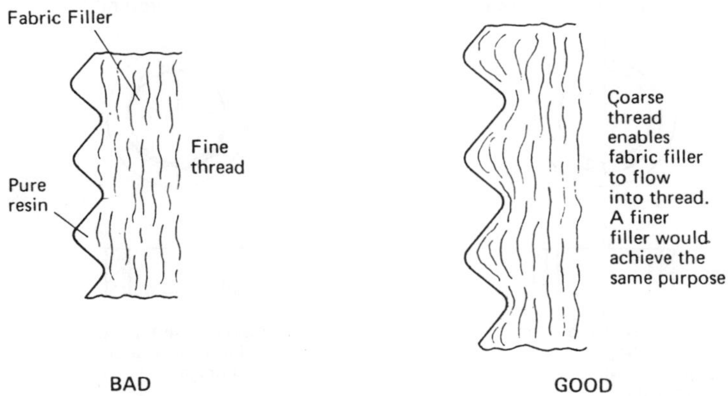

Figure 1.15 Fine screw threads moulded into fabric-filled materials are useless: the fabric filler used to strengthen the material will not flow into the fine thread. Only the resin is forced in, and consequently the thread is relatively weak.

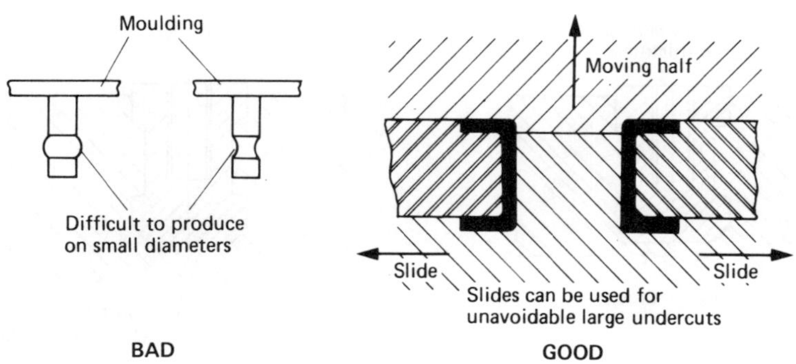

Figure 1.16 Undercuts on small diameters or projections are virtually impossible and must be eliminated.

1.5.3.3 Incorporation of moulded-in inserts

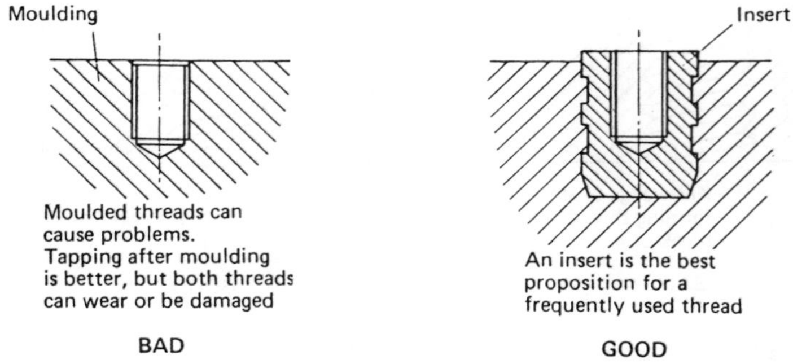

Figure 1.17 If internal threads are required, consider tapping them after moulding. On small threads, thread length should not be more than 2 × diameter. If the screw is to be removed and replaced frequently a tapped metal insert is preferable.

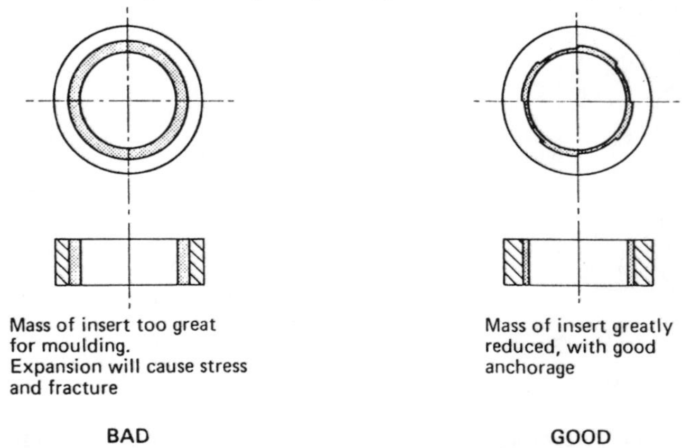

Figure 1.18 When planning the use of inserts in the moulding, consider the relative thermal expansion of plastics and the metal inserts, and expect stresses. Keep inserts as small as possible, consistent with adequate anchorage.

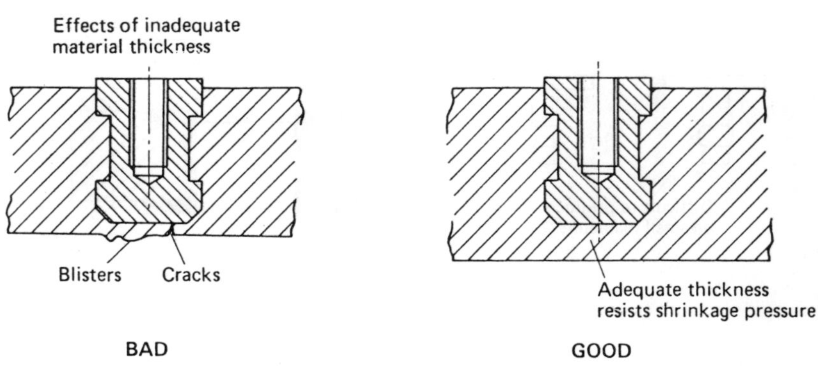

Figure 1.19 Thickness of material above a moulded-in insert must be adequate to resist shrinkage pressure. If it is too thin it will blister and/or crack.

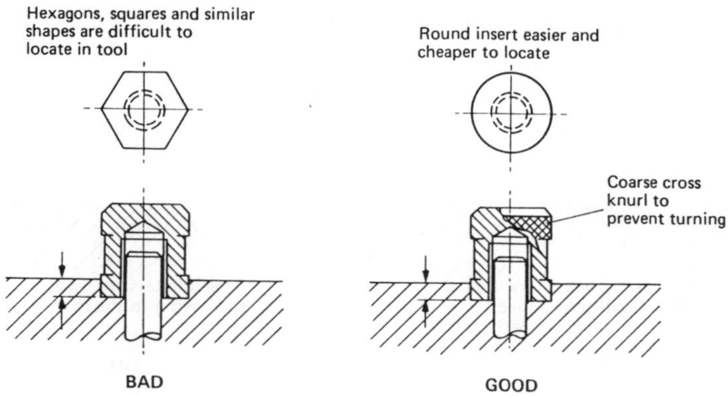

Figure 1.20 Do not call for hexagonal or irregular-shaped inserts that project from the surface of the moulding. The irregular-shaped receiving hole in the mould may be difficult to machine accurately, particularly if it is deep down.

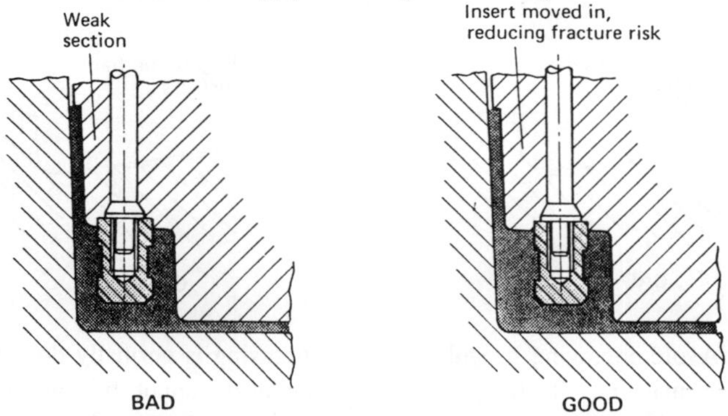

Figure 1.21 A projecting insert must not be too close to the edge of the moulded part, since the steel section would be too thin and might crack in hardening or as a result of fatigue when moulding.

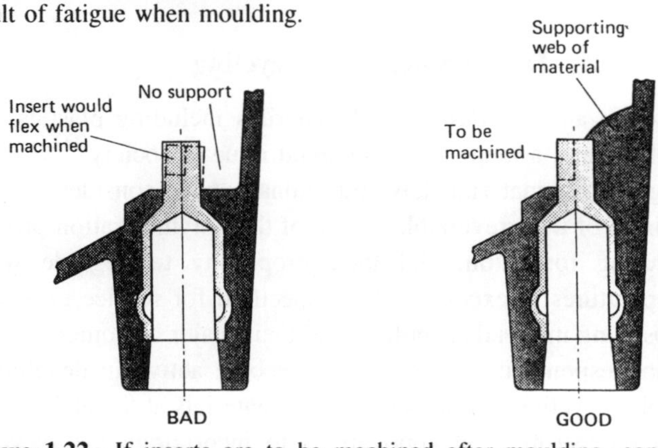

Figure 1.22 If inserts are to be machined after moulding, consider the possible heating effects and the disparity of thermal expansion, and also the stresses set up in the machining operation — these may well be transferred to the moulding and cause cracking.

1.5.3.4 Incorporation of engraved lettering

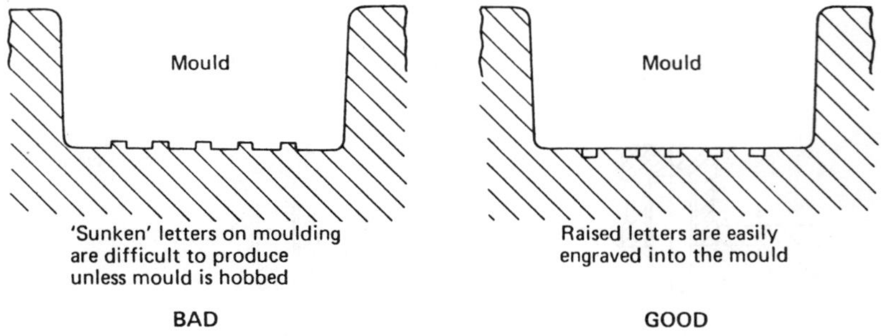

Figure 1.23 Use raised lettering wherever possible — it is easily engraved into the die. Sunken lettering requires raised features in the die, which are very difficult unless they can be 'hobbed'.

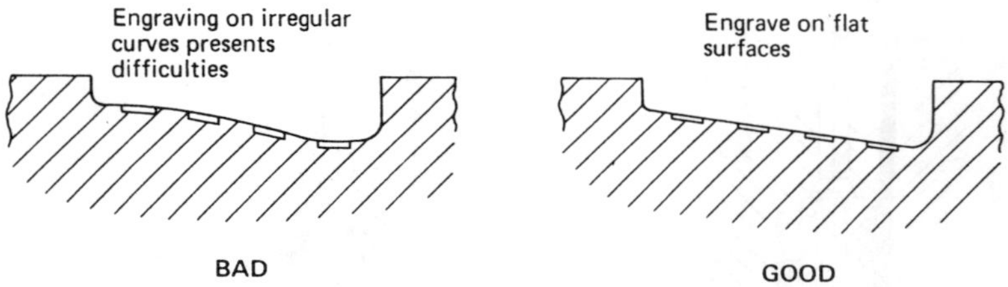

Figure 1.24 Avoid engraving irregular curves. Engraving is normally done with a pantograph-type machine. If the piece to be engraved cannot be swung on a mathematical curve, it needs to be reset for each letter — and for large letters, perhaps for each part of a letter.

1.6 Developments in recycling

The subject of waste and recycling of all materials including plastics, has become, and will remain, a major environmental issue in society.[11]

Thermosetting plastics materials have traditionally been considered 'non-recyclable' by virtue of the irreversible nature of the polymerization process they undergo during formation, and their propensity to degrade when subjected to temperatures in excess of those specified for service exposure. Since the thermosetting materials suppliers and their major customers cannot ignore their social responsibilities, they have become active in developing processes to enable recycling of components and materials at the end of their life, and also in pioneering new applications to incorporate the recycled materials. The thermoset industry is taking the firm view that without effective recycling strategies to implement, their products may not have a viable future.[12]

A recent study of the recycling problem in Western Germany has shown that plastics account for roughly 70%, or 2 million tonnes, of all waste generated annually. More than 90% of this amount is thermoplastics, which in theory is amenable to melting and re-use, but may still present problems in sorting and separation to arrive at a usable feedstock. Although thermosetting materials present the same problems in sorting and separation, and additionally require more specialized treatment to enable recycling, they will continue to be specified and used in the future. The principal reason for this is that products made from reinforced thermosetting materials continue to offer preferred and, in many cases, unique properties compared with other materials. Furthermore, the products made from thermosetting materials are often lighter, cheaper, have lower total energy requirements for manufacture than other usable materials (as mentioned previously in section 1.2), and compare very favourably with other materials in an ecobalance, or life cycle energy analysis.

Many initiatives are in progress to establish viable recycling processes, and references are provided relating to some of the more important developments.[13–15] It is, however, worthwhile mentioning one development, the ERCOM project, which serves as an example of such activities.

ERCOM is the name of a recycling plant and project established by the

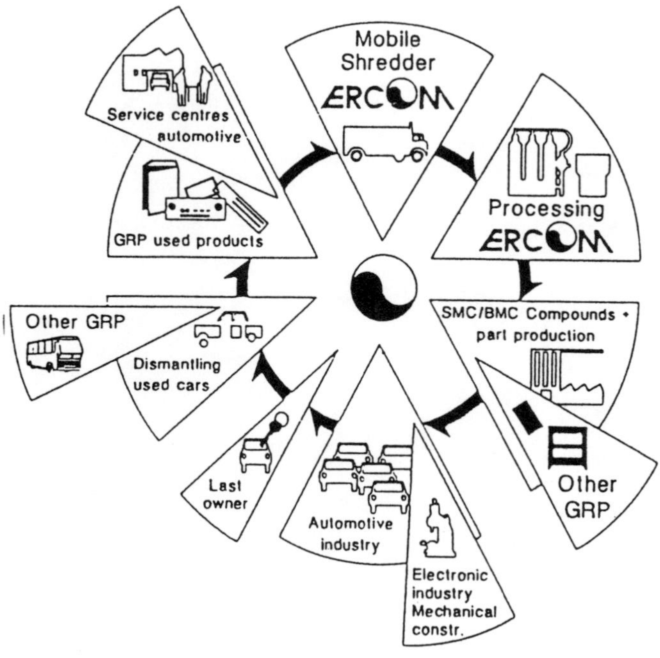

Figure 1.25 Material recycling system in the ERCOM project. Reproduced by kind permission of The British Plastics Federation from Conference Paper 20: BPF Composites Congress 1992.

German Federal government in partnership with a group of major raw materials suppliers to undertake recycling of BMC and SMC components from motor vehicles to a standardized recyclate (Fig. 1.25). Input of waste parts is from the part producer, whereas parts from dismantled cars come through a companion project PRAVDA, which is set up to establish a countrywide network of dismantling plants. The target is for 10 000 tonnes recyclate throughput from the plant per year. Components are shredded within the ERCOM plant, the usable shredded fraction is separated from other materials and then is processed into categorized fillers for BMC or SMC products.

Tests have shown that, depending on requirement for the products, 10–30% recyclate can be used in place of traditional glass fibre reinforcing materials. The used material is separated into six well-defined particle size fractions in fibre and powder form and supplied to manufacturers of new reinforced plastics.

CHAPTER 2

General characteristics and properties of thermosetting moulding materials

2.1 Polymerization reactions

The properties of plastics materials and the methods by which they may be processed are significantly affected by the chemical structure and the nature of the constituent materials from which the polymer is created.[16,17]

In simple terms, all polymers are formed by the chemical linking of small molecules, or monomers, to form very large molecules in a polymerization reaction.

The main categories of polymerization reaction are described and illustrated below, and it will be seen that they are applicable for creating thermoplastic and thermosetting polymers.[1] However, as has already been mentioned in the previous chapter, thermoplastic polymers have more or less linear molecules and therefore soften and flow when heat and pressure are applied and re-solidify on cooling. On the other hand, thermosetting resins undergo an irreversible chemical reaction or cure during the moulding process. In this process the linear organic polymer segments become cross-linked to form a rigid, infusible three-dimensional network-like structure. This reaction may be accomplished with or without the application of heat or pressure, depending on the resin type. Once cured, thermosets cannot usually be returned to the uncured state or be reshaped.

Most thermosets are processed in partially polymerized form, only the final cross-linking being carried out during the fabrication step. The main reason for this is that the polymerization is highly exothermic and may also result in shrinkage and the elimination of condensation products such as water. Carrying out as much of the polymerization as possible at an early stage minimizes the problems due to these effects.

2.1.1 Addition polymerization (sometimes referred to as free radical polymerization)

A low molecular weight (monomer) molecule possessing a double bond is induced by a suitable catalyst to break the double bond with resultant generation of free radicals which enable it to join to other similar molecules in a chain reaction mechanism.

A typical example is the formation of polystyrene.

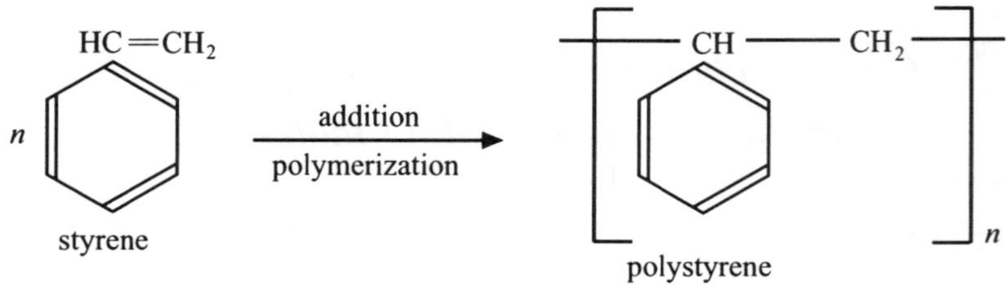

No side products are formed. This is a common mechanism for the formation of the many thermoplastic polymers.

2.1.2 Condensation polymerization (sometimes referred to as stepwise polymerization)

In its simplest form condensation polymerization differs from addition polymerization in that two different precursor species with reactive end groups react together with the elimination of a small molecule (often water) to form a long chain polymer in a stepwise condensation process. Bi-functional monomers are used to create linear polymers, while polyfunctional monomers give branched or cross-linked polymers. A typical example is the manufacture of a linear polyester from a dibasic acid and a glycol.

$$\underset{\text{dibasic acid}}{HO\vdots OCR^1CO\vdots OH} + \underset{\text{glycol}}{H\vdots OR^2O\vdots H} + \underset{\text{dibasic acid}}{HO\vdots OCR^1CO\vdots OH}$$

$$\longrightarrow \underset{\text{linear polyester}}{OCR^1COOR^2OOCR^1CO} + 2H_2O$$

The formation of a linear polyester by condensation can be used to create unsaturated polyester pre-polymers for subsequent reaction into thermosetting polyester-based resins, as described below in subsection 2.1.3. Condensation can also be used to form some nylons and the important thermosetting families of amino resins, epoxides, phenolics and silicones.

The two important classes of amino resins are the condensation products of urea and melamine with formaldehyde. The first stage of the synthesis involves an addition reaction to form methylol compounds, as indicated here.

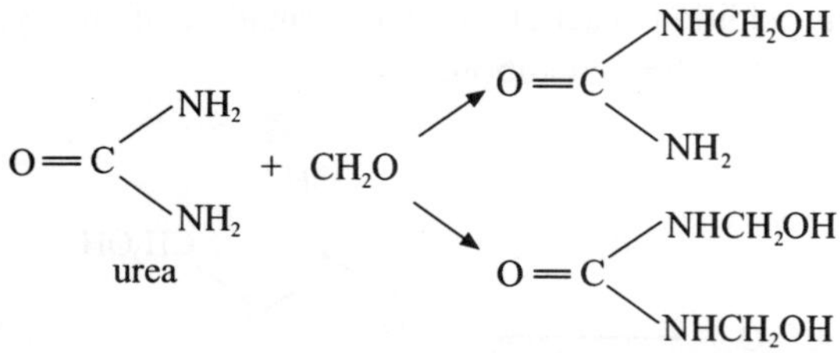

urea

mixture of mono and dimethyl urea

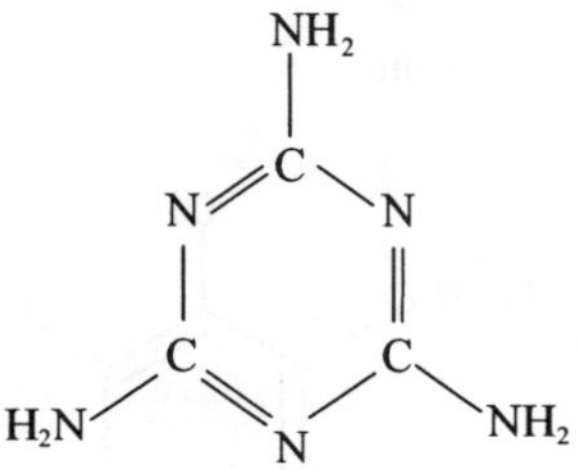

melamine, a trimer of cyanamide,
reacts similarly

A condensation reaction then follows in which methylene bridge formation occurs with the elimination of water to give structures of the following type:

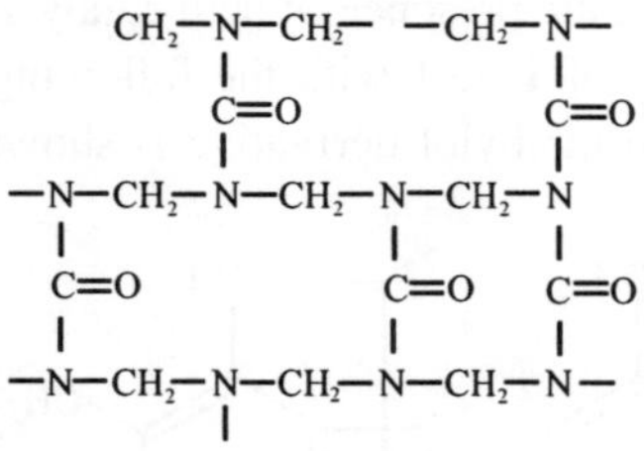

urea – formaldehyde resin

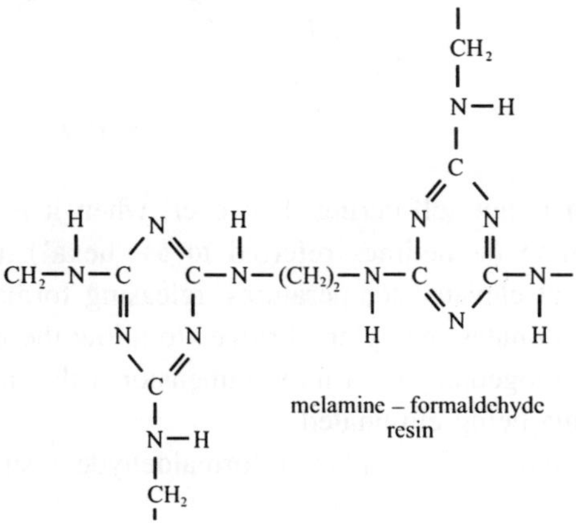

melamine – formaldehyde resin

For phenolics an addition reaction for the formation of methylols takes place under neutral or alkaline conditions.

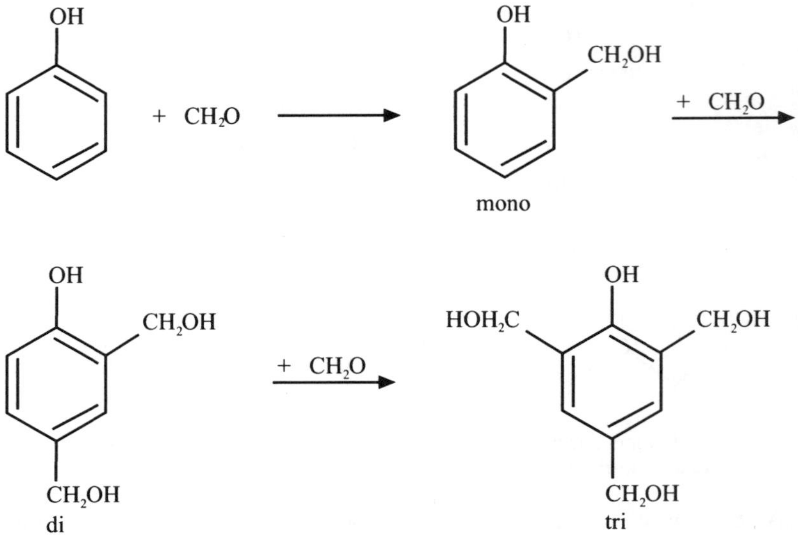

Phenolic novolak resins, most commonly used for moulding compounds, are formed by condensation in the presence of acid catalysts with a molecular ratio of phenol:formaldehyde of 1 :< 1 with the following general reaction. For simplicity only the mono methylol derivative is shown reacting.

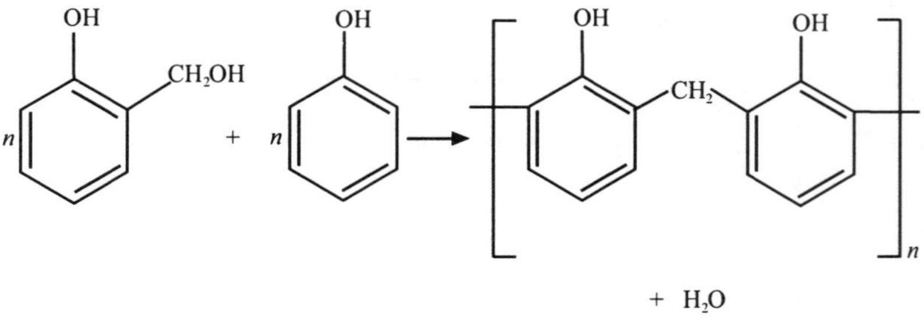

The resultant resin is not self-curing. However, when it is reacted with hexamethylenetetraamine (sometimes referred to as ‘hexa’), the ‘hexa’ is caused to decompose at elevated temperatures, releasing formaldehyde and ammonia. The ‘hexa’ donates methylene bridges to allow the novolak resin molecules to condense together into a three-dimensional thermoset network polymer, with ammonia being eliminated.

A general representation of the phenol–formaldehyde resin structure is given below:

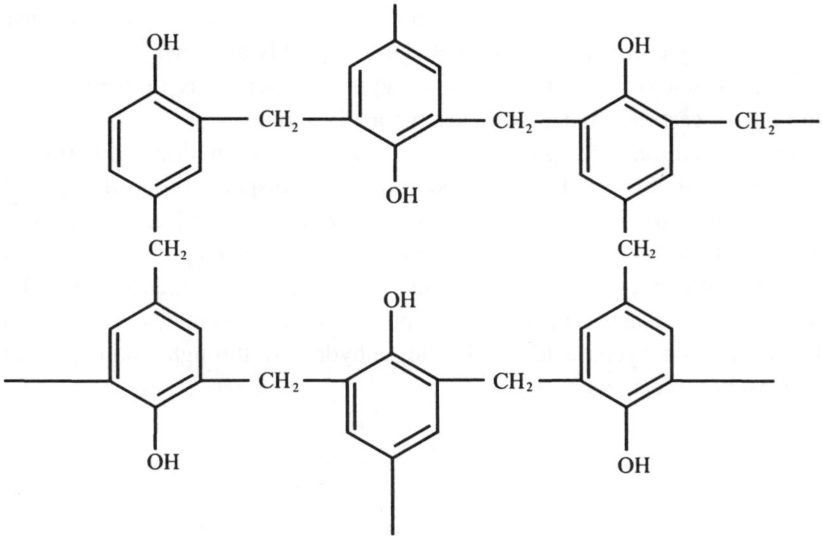

During curing the phenolic resin goes through a rubbery stage, in which it is not readily meltable and will only swell in solvents. This is sometimes referred to as resitol or 'B' stage.

Fully cross-linked resin, at the stage that melting cannot occur and the resin is insoluble in solvents, can also be referred to as resit or 'C' stage material.

The other common phenolic resins, known as resols, are formed with an excess of formaldehyde in an alkaline medium, this time with the elimination of formaldehyde and water.

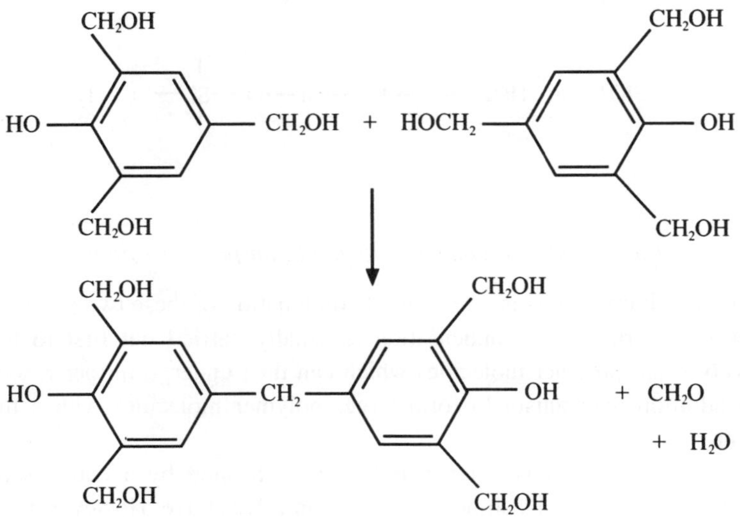

The resins formed to this stage are not stable, and spontaneously condense further, slowly at room temperature and more quickly at higher temperatures. They are often supplied in alcohol or aqueous solution as fluid resins for laminating, casting, binder and adhesive applications.

The formation of epoxide resins from epichlorhydrin and bi- or polyfunctional hydroxyl compounds such as bisphenol A (diphenylol propane) is also a condensation reaction which takes place in aqueous caustic soda. When an excess of epichlorhydrin is used an epoxy group is left on each end of the polymer chain. Such epoxy resins can then be cured by many types of material, including polyamines, polyamides, urea– and phenol–formaldehyde, acids and acid anhydrides, through coupling and condensation reactions.

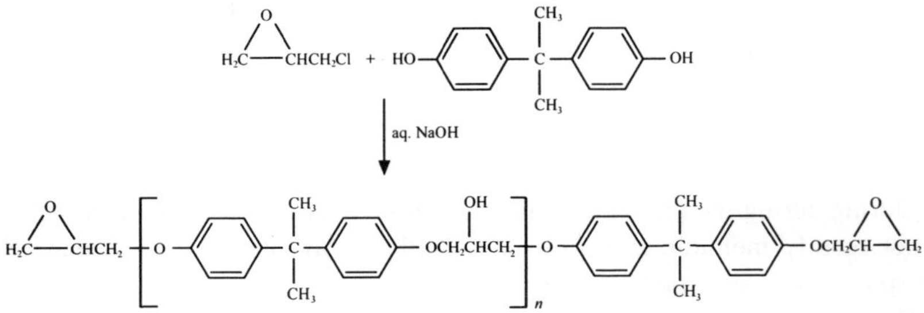

Silicone polymers are formed by a condensation reaction of silanols, which are formed from the appropriate halide or alkoxy silicone intermediates by hydrolysis.

$$\text{—SiCl} + H_2O \longrightarrow \text{—SiOH} + HCl$$

$$\text{SiOH} + \text{HOSi—} \longrightarrow \text{—Si—O—Si—} + H_2O$$

2.1.3 *Combined condensation/addition polymerization*

As the name implies, this is a sequential combination of these two processes. In this polymerization a condensation is usually carried out first to form relatively small polymer molecules which can then undergo further reaction by an addition mechanism to form larger polymer molecules with a third reactant.

An example of this is when a polyester is formed by a condensation reaction similar to the one shown in subsection 2.1.2 above. However, in this case the polyester is formed with residual unsaturated double bonds which are caused to combine with styrene in a free radical addition reaction similar to the example given previously in subsection 2.1.1, to form a thermosetting polyester–styrene copolymer as shown below.

2.1.3.1 Step 1 Condensation reaction

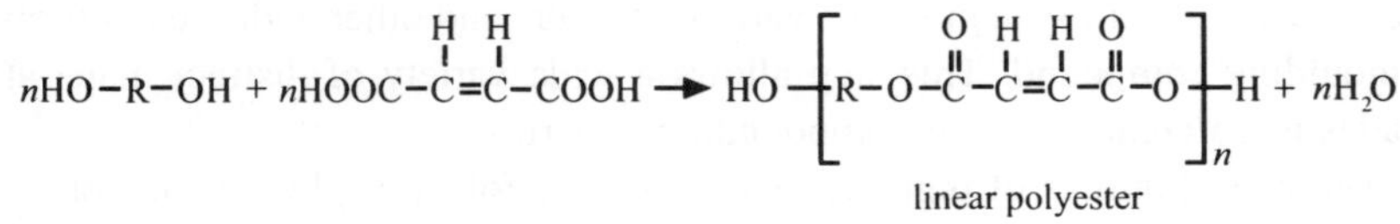

2.1.3.2 Step 2 Addition reaction (schematic illustration)

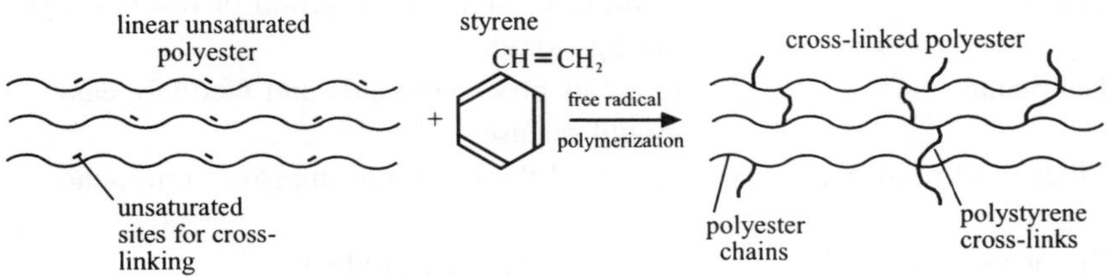

2.1.4 Rearrangement polymerization

There are two common versions of rearrangement polymerization, which are both essentially addition-type reactions.

The first occurs through a rearrangement of atoms within the two reacting monomer species, without net loss of any reacting atoms. An example of this form of rearrangement reaction is shown in the following polyurethane formation.

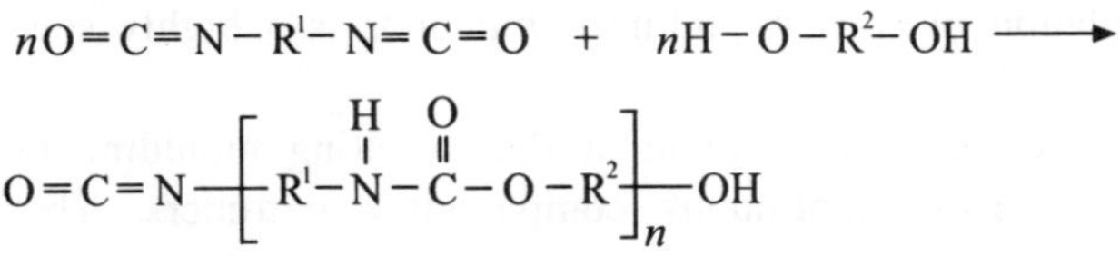

The second version occurs when a molecule composed of a ring of atoms opens up and connects with other ring-type molecules opening up in a similar fashion during the reaction. An example of a ring-opening reaction is the formation of nylon-6 from caprolactam.

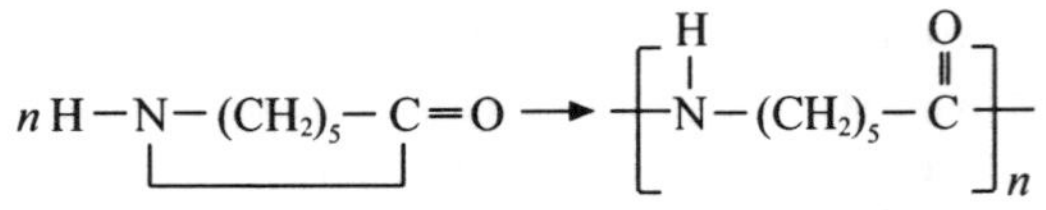

caprolactam nylon-6

2.2 Moulding compounds

It has already been mentioned that it is common practice to mix the thermosetting resin system with a variety of fillers and other additives to form a moulding compound. This step allows a wide variety of distinct material grades to be created by the compounding expert.

The main families of ingredients fall into the following broad categories.

Catalysts (initiators)	to cause full cross-linking of the moulded part
Accelerators	to promote and modify the rate of chemical reaction
Inhibitors	to prevent premature reaction of resin/catalyst during storage
Lubricants	to act as processing aids and facilitate easy mould release
Fillers and reinforcements	to extend the resin and improve compound properties
Pigments and dyes	to colour the final product

A more detailed discussion of additives is given in Chapter 4.

The manufacture of a thermosetting moulding compound is usually based on a control of the polymerization reaction to enable it to effectively become a two-stage process. The first part of the reaction is started during the compounding phase and the degree of 'pre-cure' that is allowed to occur is sufficient to assist the binding in of ingredients and granulation for ease of subsequent handling, but not so much as to restrict the ability of the compound to resoften and flow in the mould prior to final achievement of full cure. A typical example would be in the manufacture of a compound based on a phenolic novolak resin. As can be appreciated from this brief discussion, the formulation of such moulding compounds is a highly specialized and skilful process.

Today, the majority of granular thermosetting moulding materials are compounded using continuous compounding extruders. The traditional method of compounding using a two roll mill is only retained for special applications and very low volumes.

The moulding material so made can be then substantially influenced by the conditions and duration of storage prior to moulding. Different materials families with their own recommended shelf-life conditions vary from the phenolic with a shelf-life of one to two years, while some compounds employing liquid polyester resins may have a shelf-life of a few months. In all instances the higher the temperature of storage, the shorter the relative shelf-life.

Most moulding materials contain some amounts of water from either the chemical reaction or associated with the fillers. It is usual to control the water content, since the amount not only affects the processing performance of the compound, particularly flow, but also can lead to increased shrinkage from the mould and excessive gassing during moulding. Volatiles can be encouraged to escape by use of appropriate mould design and, if necessary, mould breathing techniques, as discussed in Chapters 5 and 6.

2.3 Form

Granular materials are supplied in a variety of grinds (e.g. that retained on a 60–200 mesh sieve) or in regular granulate form or, in the case of long fibre 'high strength' materials, in rather bulky irregular form.

Choice of form is sometimes possible and, for specific processing, the form is often designed to meet the particular requirements.

Compression and transfer moulding often require that the charge of material is in a handleable form and of a specific weight. This is achieved by pelleting a fixed weight (or volume) of material into a tablet of consistent density. These pellets are usually preheated to the required temperature before insertion into the mould cavity. When this is carried out in a high frequency heater, consistency of pellet weight will aid reliable control of this heating.

Where materials are required for such applications, it is usual for the moulding materials to contain a proportion of fines, that is material that aids the bonding of the pellets into consistent form. These fines should not be in excess, nor too small in grind.

In the case of injection moulding or where automatic powder feeding is required, an excess of fines may result in clogging, which is often overcome by use of vibrators or stirrers in the hopper. It is usual in such instances for the material to contain less fines, and grades are available which have been extruded and cut into regular form which eliminates fines altogether.

Thermosetting moulding materials may also be available in other forms. Notably, liquid polyester resins can be formulated with glass fibres and particular fillers to produce dough-like bulk moulding compounds (BMCs) and sheet moulding compounds (SMCs). These families and their moulding are dealt with in Chapter 7.

Additionally, for some specialist applications, again using liquid resins as a basis, reinforced materials can be formulated in rope form or as putty-like compounds.

In general terms, it can be observed that moulding routes involving powders, pellets and tablets are in decline in comparison to those involving granules and dough-like materials.

2.4 Flow and cure

A brief introduction to the relationship between flow and cure of thermosets during curing has already been given in subsection 1.1.2.

In general it can be said that the moulding engineer is usually looking for a compound with long flowing characteristics followed by an ability to provide rapid final curing for transfer and injection moulding applications. Conversely, a stiffer (short) flowing compound with rapid cure is common for compression moulding.

Flow behaviour of the plasticized thermoset material is obviously a very critical aspect of processing, which is complicated by the non-Newtonian behaviour of the compounds being processed. This usually takes the form of shear rate dependence of viscosity, which requires, at the very least, a power law model to describe behaviour over part of the spectrum of process rates. This situation is further complicated by the rheological and thermokinetic effects of the curing reaction, which can begin during this part of the process.

The mathematical treatment of this subject is very complex, and beyond the purpose of this book. Furthermore, the detailed approaches have to be tailored to the characteristics and reaction behaviour of specific families of thermosetting compound. A number of key reference works are listed for the reader who wishes to delve more deeply[18–23]. For our purposes we will continue to discuss the subject in four parts below.

2.4.1 Rheological behaviour

Rheology can be defined as the science of deformation and flow of matter.[1,17] When a stress is applied to a viscous fluid, the fluid will continue to deform until the applied stress is removed. In the case of a Newtonian fluid, the stress (τ) is directly proportional to shear rate (γ), that is:

$$\tau = \eta\dot{\gamma} \qquad [2.1]$$

where η is a constant of proportionality called the Newtonian viscosity.

The majority of polymers deviate from Newtonian behaviour. It therefore becomes necessary to allow the viscosity to be a function of shear rate to enable description of viscous flow behaviour:

$$\tau = \eta(\dot{\gamma})\,\dot{\gamma} \qquad [2.2]$$

The most common deviation for polymers is for viscosity to show a reversible decrease with increasing shear rate, termed pseudoplastic, or shear-thinning behaviour. The less common behaviour is called dilatant or shear

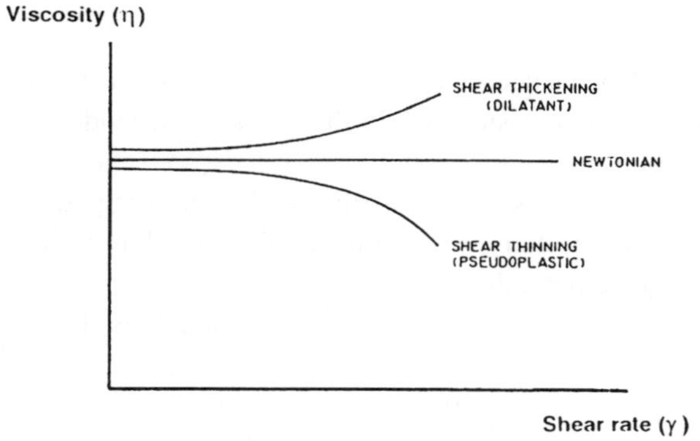

Figure 2.1 Characteristic viscosity behavioural patterns of polymers.

thickening. These characteristic behaviour patterns are illustrated in Fig. 2.1. Shear rates can range up to several thousands per second for conventional moulding processes.

Some materials, such as highly filled and/or reinforced polymers, may be capable of supporting a finite stress up to some critical yield value, above which continuous deformation occurs continuously like a viscous fluid, as illustrated in Fig. 2.2.

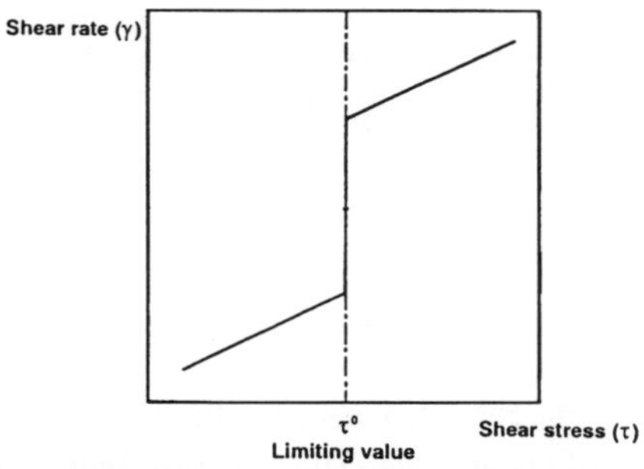

Figure 2.2 Viscosity behaviour of polymers having a critical yield value.

Various empirical, semi-empirical and theoretical relationships have been developed in order to account for the observed non-Newtonian behaviour of polymer systems. For many polymer melts, the logarithm of viscosity plotted as a function of the logarithm of strain rate shows linear behaviour over the range of strain rates encountered during processing, suggesting that a simple power law expression can be used as follows:

$$\tau = m\dot{\gamma}^n \qquad [2.3]$$

where m is a consistency index and n is the power law index, both dependent on the particular polymer and temperature. For a pseudoplastic melt $n < 1$. Leading on from Eqn 2.3 it is apparent that the viscosity is given as:

$$\eta = m\dot{\gamma}^{n-1} \quad [2.4]$$

The temperature dependence of the consistency index can be considered to follow an Arrhenius expression, while the pressure dependence follows an exponential relationship.

Although Eqn 2.4 does not properly predict limiting zero shear rate viscosity, its simplicity and small number of material parameters make it attractive for engineering calculations.

One of the most popular empirical viscosity models is the Bird–Carreau expression:

$$\frac{\eta - \eta_\alpha}{\eta_0 - \eta_\alpha} = [1 + (\lambda\dot{\gamma})^2]^{\frac{n-1}{2}} \quad [2.5]$$

where η_α is the upper limiting Newtonian viscosity and λ is a time constant. For polymer melts good agreement with experimental data can be found for $\eta_\alpha = 0$.

Since η_α is usually small compared to other quantities, this equation can be reduced to:

$$\eta = \eta_0[1 - (\lambda\dot{\gamma})^2]^{\frac{n-1}{2}} \quad [2.6]$$

Another popular empirical viscosity model is the modified Cross expression:

$$\eta = \frac{\eta_0}{1 + \left[\frac{\eta_0\dot{\gamma}}{\tau^0}\right]^{1-n}} \quad [2.7]$$

where

$$\eta_0(T) = A\exp[T_b/T] \quad [2.8]$$

η_0 is the temperature dependent zero shear rate viscosity which can be applied over a wide range of shear rates. The equation includes four material constants τ, η, A and T_b which are independent of temperature. This equation has been widely used for simulation of thermoplastics injection moulding.

Figure 2.3 gives a comparison of Bird–Carreau and Cross models with experimental viscosity data for polystyrene.

The discussion thus far has provided an introduction to the rheological behaviour of essentially thermoplastic polymers, not taking into account the added complication of the curing reaction which occurs with thermosetting materials.

For thermosets, many material properties undergo major changes as the cure reaction proceeds. However, in some instances, for example with free radical cure systems (such as BMCs), which display a low rate of cure below their cure temperature, it is often the objective to fill the mould before any

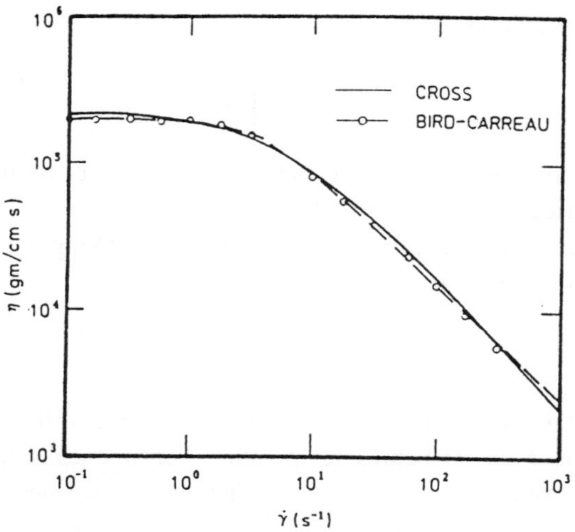

Figure 2.3 Comparison of Bird–Carreau and Cross models with experimental viscosity data (polystyrene).

significant cure can take place. In these cases rheological study can be relatively easy, and if rheometric information only is needed, it is sometimes permissible to omit the curing agent. However, this is not always the case, and, particularly in situations such as the reaction moulding process, where reactive mixing occurs prior to mould filling, it is important to consider the influence of the state of cure on shear rate dependent viscosity.

The basic relationships for viscosity dependence given previously (see Eqn 2.1 and 2.2) now have to provide for the dependence of viscosity on the state of advancement of the cross linking reaction in the following manner:

$$\eta = \eta_0 f(\dot{\gamma}) g(T) h(c) \qquad [2.9]$$

where η_0 is the zero shear rate viscosity, f is the function of shear rate dependency, $g(T)$ is the function of temperature dependence and $h(c)$ is the function relating to the state of advancement of the cross-linking process.

Using the traditional power law representation of shear rate dependence:

$$\eta = \eta(\dot{\gamma}) = \eta_0 \dot{\gamma}^{n-1}$$

and temperature dependence according to Arrhenius Law, it is possible to describe the variation of viscosity with time as follows:

$$\ln\eta(T) = C_1 + \frac{\Delta E_\eta}{RT} \qquad [2.10]$$

where ΔE_η is the activation energy of the reaction, R is the gas constant, and C_1 is a constant.

From this we derive:

$$\eta = \eta_0 \gamma^{n-1} \exp\left[\frac{\Delta E_\eta}{RT}\right] h(c) \qquad [2.11]$$

in which the function $h(c)$ for dependence on degree of cross-linking has been derived empirically by various authors.

One well used approach for modelling the cross-linking (cure) reaction has been proposed by Kamal:

$$\frac{dc}{dt} = (A + Bc^n)(1 - c)^m \qquad [2.12]$$

where c is the degree of cross-linking, t is the time, A and B are rate constants assumed to have an Arrhenius temperature dependence, and n and m are constants independent of temperature that are sometimes subject to the assumption $n + m = 2$ (second-order kinetics) to simplify calculations.

Another popular model for this function is provided by Castro and Macosko:[21]

$$h(c) = \left[\frac{C_g}{C_g - C}\right]^{A+Bc} \qquad [2.13]$$

where C_g is the degree of completeness of curing reaction at the gel point, A and B are constants depending on the polymeric reagent and $h(c)$, always ≥ 1, gives the increase in viscosity up to the point when $\eta = 0$ when the gel point (the point at which flow ceases) is reached.

Combining Eqns 2.11 and 2.13, we can arrive at a general model for the viscosity of a thermosetting system as follows:

$$\eta = \eta(\dot{\gamma}, T, C) = \eta_0 \dot{\gamma}^{n-1} \exp\left[\frac{\Delta E_\eta}{RT} \cdot \frac{C_1}{C_1 - C}\right]^{A+Bc} \qquad [2.14]$$

Derived viscosity models for thermosetting moulding compounds similar to that shown in Eqn 2.13 have been successfully used in the development of commercial computer software packages for moulding process simulation.

2.4.2 Flow effects in the mould

Orientation occurs during flow in channels, or between mould plates as molecular segments, or the larger entities often present in thermoset moulding compounds, are forced in a preferred direction.

In Fig. 2.4. we see a schematic representation of different flow fronts that can be generated. Type A shows the behaviour of a Newtonian fluid, resulting in a parabolic velocity profile, which contrasts with Type B which is typical of the non-Newtonian behaviour of many polymers. Type C shows the

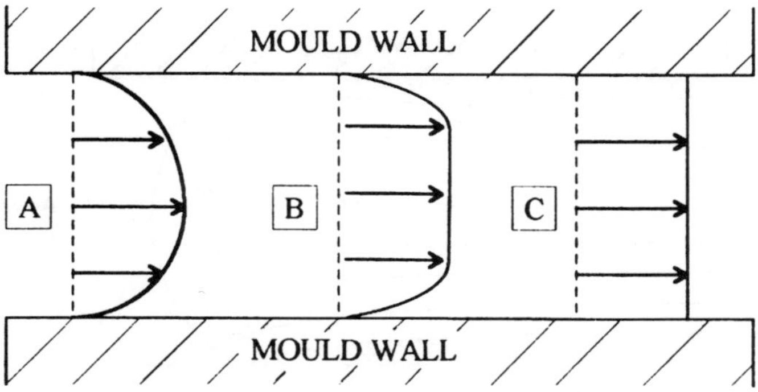

Figure 2.4 Simple channel flow between parallel plates. Type A — isothermal flow of a Newtonian fluid. Parabolic flow front. Type B — effect of non-Newtonian behaviour. Flow front is flatter in the centre and sharper at the edges. Type C — plug flow, the extreme case of non-Newtonian flow. Often approached with highly filled BMCs.

extreme case of plug flow behaviour more common with highly filled moulding compounds such as BMCs. It is worth noting that Type A and Type B behaviour encourage higher levels of orientation near the mould wall than in the centre of the channel.

It can be seen that the movement of the moulding compound closest to the mould wall is slower than in the centre, thus resulting in marked velocity differentials. Some molecular relaxation may occur when movement ceases, causing the polymer to attempt to return to a disorderly molecular arrangement. This is, however, severely hampered in the case of thermosets by the presence of large amounts of filler and reinforcement in the compound and also by the onset of cross-linking.

Now considering Fig. 2.5, we see the onset of curing, which starts to create a solid layer at the mould wall which restricts the passage for subsequent material flow, and also tends to promote fountain flow behaviour at the flow front. The narrowing of the flow channel resulting from the build up of the solid layer has the effect of moving the area of highest orientation in the compound nearer to the centre of the channel. This is a more important consideration with thermoplastics, where the solidified surface layer tends to be much thicker than with thermosets. It is also the case that fountain flow effects are often less pronounced with thermosets than with thermoplastics.

This model is also useful in understanding how flow in the mould can result in a three-layer structure in the moulding. When fibrous fillers are present, their final orientation is heavily influenced by this effect. It is normal to find that the fibres will be predominantly aligned in the flow direction in the surface layers (which solidify during mould filling), while the central

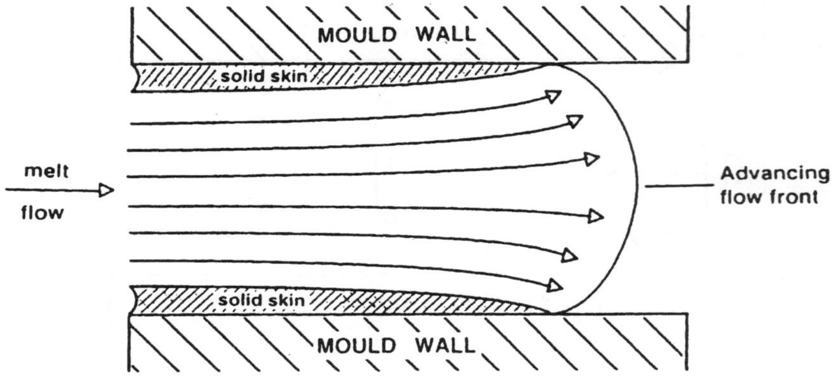

Figure 2.5 Solidification (curing) effects during mould filling.

layer is likely to have predominantly transverse fibre alignment due to the influence of divergent flow of the material from the mould gate. Once again, it should be noted that the surface layers, with fibres aligned in the flow direction, can be very thin with some thermosets such as BMCs.

It should be noted that there is a significant difference in the behaviour of thermoplastics and thermosets in these circumstances. When molten thermoplastic passes between the mould walls during mould filling it is losing heat steadily, and therefore the melt viscosity is constantly increasing. In the case of a thermoset moulding compound, it must be remembered that the viscosity characteristic of a thermosetting resin is to exhibit a viscosity reduction during the first phase of flow (into the mould), followed by a viscosity increase with the onset of the curing reaction.

The thickness of the solid (cured) skin at the mould wall increases with the degree of cure, causing a restriction for the flowing melt. In practice it is important to design the moulded part to have adequate section thicknesses to allow complete filling of the mould cavity before flow restriction reaches a limiting value.

The presence of fillers and reinforcements in the compound also leads to high entry pressures in nozzles, gates and regions of similar geometry (i.e. convergent flow). It is therefore usual to make a correction to viscosity calculations for heavily filled materials to allow for die entry and exit pressure effects, sometimes referred to as 'end effects'. The method of measurement and calculation of these end effects is treated in subsection 2.4.4.

For thermoplastic injection moulding it has become increasingly popular to use the so-called 'P-V-T' diagrams to predict the volume change (shrinkage) effect within the mould. These diagrams attempt to depict the dependency of the volumetric shrinkage of the moulded part on moulding pressures and temperatures, making use of the equation of state relating pressure P, temperature T and specific volume V. However, in the case of thermosetting injection moulding there is scarce evidence to indicate that sound data exists for such calculations, and furthermore, that such calculations would be reliable in a process where a chemical reaction is now superimposed.

2.4.3 Cure behaviour

The basic mechanisms governing the cross-linking behaviour of the main thermosetting compounds have already been discussed earlier in this chapter. Mention has also been made at the start of this section of the usual practice of tailoring the reaction chemistry and thermokinetics to achieve different curing speeds for different applications.

To assess the curing behaviour of the material, it is normal to consider three elements of the reaction: (i) time taken for the reaction to start, (ii) time taken for the peak reaction exotherm to be reached and (iii) temperature achieved at the peak exotherm point. It can be difficult to make detailed measurements of these three elements in a single test. Actual tests used, such as the SPE cure test, can be informative despite containing a considerable level of empiricism. In addition it is often useful to have a simple measure of the combined flow and cure behaviour of the compound using a standard spiral flow mould. Both of these tests are described in the next section.

2.4.4 Measurement of rheological and cure characteristics

The transport equations and constitutive relationships contain material functions or properties which must generally be determined experimentally. In addition, most polymers are subject to elevated pressures during processing. For example, pressures ranging up to $50\,\mathrm{MN\,m^{-2}}$ are typically encountered in polymer moulding operations. Although most material properties do not vary significantly over moderate ranges of pressure, substantial changes may occur for marked elevations in pressure. Thus, characterization of material properties should be made over the pressure range for processing if the expected pressure levels are likely to be pronounced.

Viscosity measurements are commonly made using a variety of rheometers developed to allow investigation of shear rate and/or cure effects. The two most common types of rheometer are the capillary version and the parallel plate. Full description of these devices can be found in specialist works elsewhere.[24,25]

The determination of the rheological behaviour of thermosetting materials is made difficult by the fact that the resin undergoes reaction during the measurement period. There are many tests such as the ASTM Cup Flow, the Orofice Flow and the Spiral Flow developed over the years to characterize the rheological behaviour of thermosetting compounds, as well as more elaborate tests and equipments such as the Brabender Torque Rheometer and the Monsanto Disc Rheometer. However, all these suffer from the limitation of not providing information on the fundamental rheological properties that can be used in the mathematical modelling of the thermosetting moulding process. They can, however, serve a useful purpose in routine testing of the quality and performance of the compound, as discussed later in this section.

In order to obtain a more fundamental understanding of the rheological behaviour of reactive systems, a variety of methods can be employed. The most commonly employed technique involves shearing a sample of the thermosetting material using cone-and-plate fixtures. The two most widely used commercial instruments are the Weissenberg Rheogoniometer and the Rheometrics Mechanical Spectrometer. These instruments have the capability for temperature control by means of an environmental chamber which surrounds the test fixture. In addition to determining the viscosity as a function of time at a particular temperature and shear rate, these instruments can also yield information with regard to the first normal stress difference by measuring the normal force or thrust on the platens. The cone-and-plate geometry restricts the measurements to relatively low shear rates, and also the time required for the sample to attain thermal equilibrium limits meaningful use to systems that do not polymerize rapidly.

In order to characterize the material at high shear rates (around $1 \times 10^3\, s^{-1}$), a capilliary rheometer can be used. In general terms, capillary rheometers will yield useful data on the shear rate model parameters, as well as evaluation of slip, yield stress and sample fracture due to filler loading of the moulding compound. They are, however, unsuitable for use with solid compounds, or compounds where cure begins when they are heated to the measuring temperature (unless the curing system has been left out).

In some cases, workers have conveniently linked capillary rheometers to thermosetting injection moulding machines, fitting the device into the platen area, to receive the melt processed by the machine in the normal manner. Usually the rheometer is capable of incorporating a range of interchangeable capillaries of difference *L/D* ratio. Viscosity measurements can be made at varying temperatures and apparent shear rates for the different capillary *L/D* ratios. As previously mentioned, it is often necessary to correct for pressure difference 'end effects', since in some cases these can amount to half of the total injection pressure in regions of strong convergence such as nozzles and mould gates.

From measurements taken with capillaries of different length, but having the same diameter and entry geometry, it is possible to plot total pressure against die *L/R* ratio. This plot, called a Bagley Plot, is illustrated in Fig. 2.6,

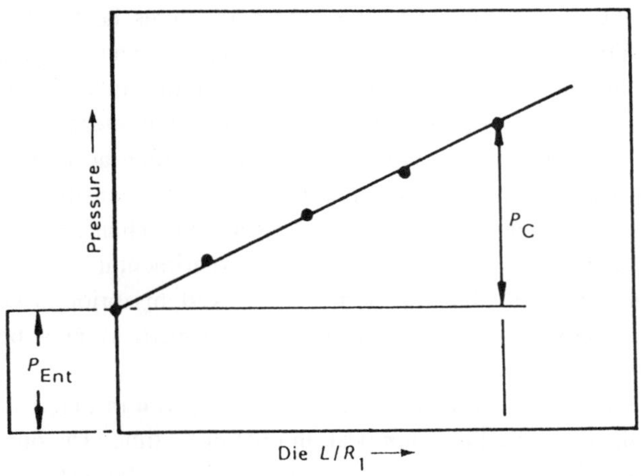

Figure 2.6 Example of a Bagley Plot for establishing die entry pressure losses.

in which the extrapolation of the plot to zero *L/R* ratio yields the value of pressure losses.[26,27]

The study of thermoset cure and kinetics requires knowledge of the heat of reaction and variations of the degree of cure with the thermal history of the material. This can be approached by a variety of experimental techniques including chemical analysis to measure the quantities of reactive species present in the system at any given moment during the reaction.[28] Instrumental analytical techniques, such as differential thermal analysis (DTA), differential scanning calorimetry (DSC) and infrared spectrometry can also be conveniently used. DTA and DSC are now becoming widely used techniques.

The study of reaction kinetics by means of differential thermal analysis (DTA) involves the comparison of the temperature at the centre of the sample with the temperature of an inert reference material subjected to the same thermal regime.

Differential scanning calorimetry (DSC) is based on the principle of differential enthalpic analysis and involves providing heat to both the sample and to an inert reference sample at a varying rate so as to maintain their temperatures equal. The calorimeter may be employed in either an isothermal or non-isothermal scanning mode. The sample can be held constant in isothermal mode or varied with time in non-isothermal mode. The endothermic or exothermic heat of reaction can, therefore, be measured as a function of time or sample temperature, depending on the mode of operation. By assuming the rate of heat evolution is directly proportional to the rate of reaction, calorimetry can be used to provide information on the kinetic behaviour of the system.

Although DSC is the only technique which measures heat input/output directly, both DSC and DTA seem to be used almost interchangeably for characterizing thermosets. The most accurate temperature measurements are given by DTA, while DSC gives most accurate heat flow measurements.

The curing reaction of a thermosetting polymer is often accompanied by changes in a variety of physical properties such as density, electrical conductivity, refractive index, viscosity and dynamic mechanical properties. Certain physical properties (e.g. viscosity) or experimental methods (e.g. dilatometry) are more suitable for thermosetting systems prior to gelation, while other methods (e.g. dynamic mechanical properties) may be more suitable for the phase after gelation.

A representative graphical result of a DSC test is given in Fig. 2.7, which clearly shows the quantified reaction peak in relation to time. The output can be used to define the heat of reaction (ΔH_r), the specific heat (C_p) and also assist in calculation of the Arrhenius equation kinetic functions.

The routine assessment of flow and cure can be examined in a number of ways. For a simple and easily executed empirical test suitable for routine quality control in production, it is possible to use a spiral flow test in which material is forced into a spiral channel in a heated mould under standardized conditions. The length of flow in the spiral to the point of cure provides a numeric measure of the flow/cure characteristic of the compound.

Disc flow and cup flow tests are alternatives more often used to assess the characteristics of compression moulding materials. With disc flow, the unrestricted distance flowed by the curing material between the two parallel

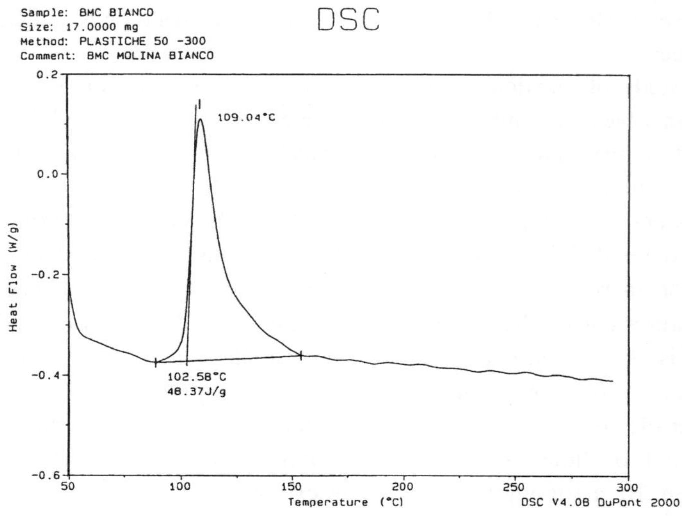

Figure 2.7 Typical graphical output of a DSC test for a BMC compound.

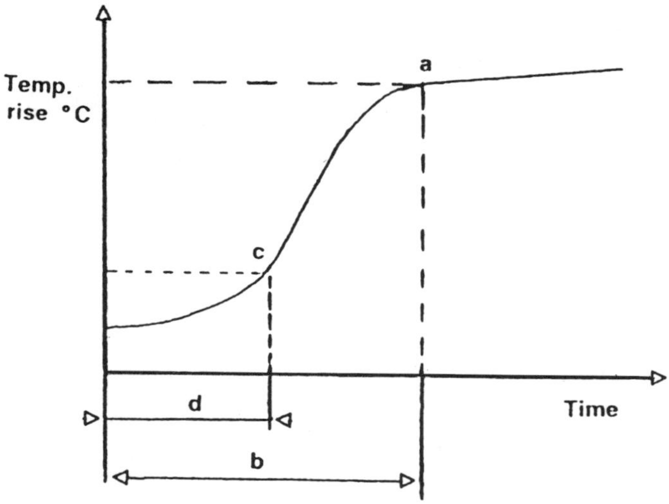

Figure 2.8 Schematic version of an SPE test result where a = peak exotherm temperature, b = time to peak exotherm (curing time), c = gel point, d = time to gel point.

plates of the mould is measured (or alternatively the final thickness). With cup flow it is usual to measure the time taken to close the mould, as a measure of flow, plus the time taken to full cure, as defined by elimination of any tendency to blistering in a deliberately designed thick section in the base of the cup.

The routine quality control cup flow and disc flow tests are becoming increasingly substituted by more sophisticated and instrumented moulding tests such as the Society of Plastics Engineers (SPE) cure test, which relies on following the cure reaction temperature in relation to time, and calculating initiation (gel) time, peak exotherm temperature and time to peak exotherm. A schematic version of an SPE test graph is shown in Fig. 2.8.

Alternatively, for some materials such as phenolics, standard chemical test procedures exist to measure the extractable (uncured) resin portion of a moulding.

Readers interested in more detailed information on standard test methods for material property characterization can find more information in relevant standard works.[16,29]

It is fair to comment that the whole subject of thermoset materials property testing and the reliability and usefulness of data obtained by different test methods and techniques still remains one of the most contentious aspects of thermosetting materials technology. It is to be hoped that further progress will be made in test method and test method standardization in the years to come.

CHAPTER 3

Properties, applications and processing of the main families of thermosetting moulding materials

3.1 Introduction

Having dealt with the basic characteristics of thermosetting moulding materials in the previous two chapters, we now look at the characteristics of the main families on an individual basis. Following the descriptive assessments, two tables of property data have been included for general guidance. Table 3.2 illustrates the typical physical and chemical properties for the most important commercially available grades, while Table 3.3 presents general guidelines to assist in their processing.

The information in this chapter has been collected and refined from a wide range of commercial literature, supported by reference to pertinent plastics handbooks.[16,30] While these tables provide a good overall summary and comparison of the main families, it is advisable to consult individual manufacturers' materials data sheets for precise recommendations.

3.2 Main families

ALLYLICS

Commonly referred to as DAP or DAIP materials.

Chemical family. Diallyl phthalate (DAP) or diallyl isophthalate (DAIP) will polymerize on heating to form an allyl resin. DAP can be used as a solvent monomer, and enables the blending of fillers, peroxide catalyst, etc. to give moulding powders that will cure by free radical chain polymerization.

DAP materials are sometimes referred to as '*ortho*', and DIAP materials as '*meta*'. DIAPs are usually recommended for higher temperature applications.

Form supplied

Granular	automatic processing
Long glass fibre flake	higher strength, compression
Short glass fibre flake	higher strength, compression
Nodular	higher strength, transfer

Some or all of the glass fibre may be replaced by mineral fillers to cheapen the compound and/or give easier processing behaviour. Organic fibres (e.g. nylon or terylene) will give lower density compounds.

Colourability. Wide colour range possible, since resin is water white.

Characteristic properties

Strengths	Excellent electrical insulation even after long hot humid exposure Good arc and tracking resistance Good dimensional accuracy and stability Excellent heat resistance to +200°C maximum Good chemical resistance Non-corrosive in electrical applications High gloss, hard surfaces
Weaknesses	High cost can limit application Some grades not resistant to alkalis

Processability. Compounds available for compression, transfer and injection moulding. Little or no volatiles are produced during moulding. Long and short fibre flake materials are normally compression moulded, while nodular materials are suitable for transfer or injection. Moulding of complex parts with close tolerances may require cooling jigs. For optimum properties, post cure mouldings at 150–190 °C for 2–16 hours.

Typical applications. Particularly suitable for high quality electrical and electronic components subject to extreme environmental conditions, for example connectors, switch and relay parts, terminal boards and potentiometer housings.

High price restricts applications in automotive and domestic electrical products in favour of polyester alkyds.

Allylic based prepregs are used to make lightweight intricate parts such as aircraft radomes and printed circuit boards. Clear resins are also used in casting applications, while DAP prepolymer can also be used in decorative laminates.

AMINOS

Chemical family. Amino resins are formed by the condensation reaction of formaldehyde with compounds containing amino groups, with the elimination of water. The most common examples are urea–formaldehyde (UF) and melamine–formaldehyde (MF). Co-reacted moulding compounds of phenol and melamine (called melamine phenolics) are also produced.

The heat resistance of UF resins is lower than MF resins, and their water absorption is also higher.

Form supplied. Moulding compounds fall into two main categories:

Alpha cellulose filled	Best colour range and surface appearance for decorative applications
Wood, flour, glass or mineral filled	Industrial applications where colour is not important

Amino resins are also widely used for laminating, wood adhesives, paper and textile treatments.

Colourability. Wide colour range possible, since resin is water white.

Characteristic properties

Strengths	Excellent heat resistance/insulation Hard, rigid, low deformation under load Excellent electrical insulation Fire and flame resistant Good dimensional stability Resistant to common organic solvents, weak acids and alkalis Resistant to scratching and abrasion Relatively cheap
Weaknesses	Mould and after-moulding shrinkage high Prone to cracking after severe wet/dry cycling Unsuitable for outdoor exposure Tendency to brittleness at low temperatures Colour changes at higher temperatures Not resistant to strong acids and alkalis

Processability. Moulding compounds suitable for compression, transfer and injection. The high polarity of these compounds makes them very suitable for high frequency preheating prior to compression or transfer moulding. During curing, relatively large amounts of volatiles are produced. Because of this, adequate venting of the mould is necessary, and chromium plating is often used to restrict staining and corrosion of mould surfaces.

Moulded MF parts may be decorated with MF resin impregnated paper overlays during moulding.

Mouldings made from amino compounds may be prone to some after-moulding shrinkage (of the order of 0.5%). When necessary, this can be brought out by a suitable post moulding stoving operation.

Typical applications. Primary applications for alpha cellulose filled UFs are in electrical wiring devices (circuit breakers, wall plates, receptacles), cosmetic closures, electric blanket control housings, buttons, toilet seats, housings, knobs and handles.

MF alpha cellulose materials are used extensively for moulded dinnerware, as well as buttons, ashtrays, utensil handles, lavatory bowls and appliance components.

Wood, glass and mineral filled compounds are used for a wide variety of industrial and military electrical parts.

Melamine phenolics gain some of the benefits of improved mechanical properties and superior heat resistance from the phenolic element and are therefore being increasingly used in more demanding electrical and mechanical applications.

EPOXY

Chemical family. Commercial epoxy resins are compounds, or mixtures of compounds, containing epoxide groups which cross-link via an addition polymerization reaction typically involving a polyfunctional amine, with no volatile by-products produced. The cured compounds resist many forms of chemical attack, and are very thermally stable. The most widely used resins are based on bisphenol A/epichlorhydrin.

Form supplied. Resins are used in two forms:

Liquid compounds	Potting, encapsulation, laminating, composite fabrications
Solid compounds	Compression, transfer and injection moulding

Colourability. Although epoxy resins are generally clear and straw coloured, the range of colours possible is somewhat limited in practice to black, red, brown, blue and green.

Characteristic properties

Strengths	Excellent mechanical properties Low power factor and high dielectric strength Excellent heat resistance to 220°C Excellent chemical resistance Good adhesion Good retained properties in demanding environments Excellent dimensional accuracy and stability Low mould shrinkage and internal stresses

	No volatiles/gases produced during moulding Good flow, low moulding pressures Excellent for insert moulding
Weaknesses	Demoulding rigidity rather low Relatively expensive Not resistant to some polar organic solvents Relatively short shelf-life

Processability. Epoxy moulding compounds are available for compression, transfer and injection moulding. Their low viscosity and easy flow characteristics enable low moulding pressures to be used and are particularly good for thin sections and mouldings with delicate inserts. The rapid cure reaction, being an addition polymerization, also enables thick sections to be moulded economically without problems of voids and entrapped gases.

The shelf-life of some compounds is limited. The wide range of fabrication and high performance composite moulding techniques possible for liquid epoxy resins falls outside the scope of this book.

Typical applications. Epoxy moulding materials are extensively used in the electrical and electronic industries for high performance applications because of their dielectric properties, low shrinkage, good adhesion and ability to retain properties in demanding environmental situations.

Typical applications are high voltage insulators, bushings, switchgear, printed circuit boards, terminal blocks and relays.

Liquid and granular grades are available and widely used for encapsulation of electrical and electronic components, usually by transfer moulding.

In mechanical engineering applications epoxy resins are found in pump vanes, rotor shaft insulations, and pump and valve parts.

PHENOLICS

Chemical family. The majority of phenolic moulding materials use phenolic novolak resins which are produced from phenol and formaldehyde by a condensation reaction involving hexamethylenetetraamine as the cross-linking agent. Ammonia is released during this reaction. Recent development work has focused on use of thermoplastic toughening additives to reduce brittleness in the cured product.[33]

Phenolic resins are inherently flame retardant and chemically resistant.

Form supplied. The range of compounds is very wide, resulting from their great popularity over many years. Properties achieved are correspondingly very diverse. The main fillers and resultant families are:

Wood flour	General purpose, low cost
Cellulose fibres	Improved toughness
Mica	Electrical
Glass or glass/mineral	High strength

Colourability. Very restricted. Usually only dark browns or black.

Characteristic properties

Strengths	Very good mechanical properties Maximum service temperature to 260°C with some grades Good all round electrical properties Hard and rigid Excellent dimensional stability Good chemical resistance GF phenolics have very good strength in thin sections Fire and flame resistant Excellent price/performance balance
Weaknesses	Restricted colour range Impurity levels too high for some electronic applications Not electrical tracking resistant

Processability. Widely used for compression, transfer and injection moulding. Not suitable for electronic encapsulation. Dielectric preheating of granules or pellets materials commonly used prior to compression or transfer moulding. This has the additional benefit of eliminating any traces of moisture from the compound that could affect the electrical properties of the moulded component.

Glass fibre and glass fibre/mineral reinforced grades can cause significant abrasive wear in moulding machines and moulds.

Typical applications. The range of applications is vast. The main application families are:

Electrical	Plugs, sockets, switch and fuse gear covers, housings, distribution boards, etc.
Automotive	Brake servo components, ignition coils, distributors, starter motors, alternators, solenoids, small motors and switchgear

Domestic appliances	Kettle handles, knobs, handles, water heater controls, castors, motor housings, iron handles, washing machine components, cookware handles

GRANULAR POLYESTER MOULDING COMPOUNDS

Sometimes referred to as granular moulding compounds (GMC), granular polyesters or polyester alkyds.

Chemical family. These moulding compounds are most commonly based on use of an unsaturated polyester resin formed by condensation of a dibasic acid such as phthalic or maleic anhydride and a glycol such as propylene glycol. The resin is then blended with a polymerizing non-volatile solvent vinyl monomer such as diallyl phthalate which is solid at room temperature.

The term 'ester' is given to reaction products of an acid and an alcohol. In the past, the term 'alkyd' has also been applied to some ester type compounds ('al' from alcohol, and 'kyd' from acid). A further traditional name 'glyptal' has also been used for products from the reaction of glycerol and phthalic anhydride.

The term alkyd is not nowadays encouraged, or widely used, the preference being to classify such materials as granular polyester moulding compounds.

Form supplied. A wide range of grades is possible, due to the different resin combinations available, plus the use of alternative filler types (mineral and/or glass fibre being the most common choices). Sometimes, unsaturated diallyl phthalate and vinyltoluene monomers are added to formulations to promote flow, and extend flow life.

The free radical initiated polymerization, together with copolymerization with added unsaturated monomers, often results in fast curing compounds, that are becoming increasingly important.

Pellets or granules are supplied in bag format, suitable for normal or automatic hopper feeding. Materials should be stored in cool, dry conditions, when a storage life of around 6 months is normal. For higher strength applications materials can be obtained in flake form to preserve fibre length.

Colourability. A wide range of colours is possible, including opaque light shades.

Characteristic properties

Strengths	Electrical insulation resistance Good dimensional stability; little or no after

	shrinkage No evaporation of volatiles and/or gases High heat distortion temperature More rigid but more brittle than phenolics Good surface finish; harder than BMC and SMC Low moisture absorption at room temperature Self extinguishing Low moulding pressures Fast curing
Weaknesses	Tendency to brittleness Mechanical properties reduced at high temperatures Resistance to boiling water and high humidity Relatively high moulding shrinkage; risk of cracking around inserts Electrical properties not maintained in extreme conditions

Processability. Compression, transfer and injection moulding possible. Compression moulding preferable for long glass fibre flake grades to maintain mechanical properties. Ease of moulding is generally good, although care is required in maintaining close temperature control. Low temperatures can result in incomplete cure. Putty forms are often used for electronic encapsulation. Abrasive mould wear can be similar to BMC/SMCs.

Mould temperature is of considerable importance, and low temperatures can result in incomplete cure, lower physical and electrical properties and poor dimensional stability. High thermal conductivity of most grades, plus resin reactivity, leads to very fast curing.

Granular polyester compounds, by virtue of their pelleted form, are more popular than BMCs for moulding large numbers of relatively small components.

Typical applications

Automotive ignition components
Precision mouldings for electrical and electronic industries
Switchgear and switch housings
Other electrical housings, connector boxes
Microwave cookware containers

POLYESTER BULK AND SHEET MOULDING COMPOUNDS

Thermosetting unsaturated liquid polyester resins are used as the basis of two main families of moulding compounds:

1. Polyester bulk moulding compounds (BMCs).
2. Sheet moulding compounds (SMCs).

Each family will now be treated separately. In addition, because of the growing importance of BMC and SMC materials, their processing and application are dealt with in greater depth in Chapter 7. In subsection 7.1.7 a brief description is provided of two new variants of polyester moulding compounds, namely thick moulding compounds (TMCs) and continuously impregnated compounds (CICs). For completeness, initial property data for these materials is provided in Table 3.2 at the end of this chapter.

The choice of polyester resin has an important impact on the compound characteristics for specific end-use applications, as shown in Table 3.1.

Table 3.1 Effect of different polyester resins on compound characteristics

Type	Examples	Characteristics
Unsaturated anhydrides and dibasic acids	Maleic anhydride	Low cost, moderate heat resistance
	Fumaric acid	High reactivity, improved heat resistance
Saturated anhydrides and dibasic acids	Orthophthalic anhydride	Lowest cost, moderate heat resistance
	Isophthalic acid	High strength, better chemical resistance
	Adipic, azelaic and sebacic acids	Flexibility and toughness
	Chlorendic anhydride	Flame retardance
	Terephthalic acid	High strength, high heat resistance
	Tetrachlorophthalic anhydride	Flame retardance
Glycols	Propylene glycol	Low cost, flexible, good water resistance
	Dipropylene glycol	Flexibility and toughness
	Ethylene glycol	High heat resistance and strength, low cost
	Diethylene glycol	Higher toughness and impact strength
	Hydrogenated bisphenol A	Corrosion resistance, flexibility and strength
	Tetrabromobisphenol A	Flame resistance

POLYESTER BULK MOULDING COMPOUNDS (BMCs)

Sometimes referred to as polyester dough moulding compounds (DMCs).

Chemical family. The starting point for these compounds is a liquid unsaturated polyester resin formed in a condensation reaction between a dibasic acid (typically phthalic or maleic anhydride) and a dibasic alcohol (typically propylene glycol). A liquid vinyl monomer (styrene) is used as a solvent/viscosity modifier, and then the filler, glass reinforcement, inhibitors and free radical catalyst (typically peroxide) are added. Curing of the composite is by free radical addition, also involving the styrene monomer (see Chapter 2).

In an important modification to the technology, a thermoplastic powder (such as polystyrene, polyvinyl acetate or polymethyl methacrylate) may be added to the resin. While it dissolves in the liquid resin, it separates out as a distinct phase during the cross linking process. At this stage the phase separation counteracts the normal shrinkage processes and provides what is commonly referred to as zero (or low) shrinkage behaviour.

It is common practice for the resin to be chemically thickened by addition of magnesium oxide or calcium oxide, which modifies flow characteristics during moulding and improves the surface finish of the final component. The thickening process (maturation) takes several hours to reach completion.

Form supplied. Since a liquid polyester resin is mixed with inorganic fillers and reinforcing fibres (usually glass) in a Z-blade or ploughshare mixer, the compound has the consistency of a bread-like sticky dough. This is supplied wrapped in PE/PA laminated bags to restrict styrene evaporation inside a box or keg. If containers are left open styrene will evaporate and the compound will harden and become useless. Most compounds have a storage life of 3 months at 20°C. Polythene gloves are required for handling.

Colourability. A wide range of colours is available in standard grades. However, pigmentation of the zero shrinkage grades can prove difficult because of uneven take up of dyes and pigments during the two-phase resin separation.

Characteristic properties

Strengths	Very easy flow and low moulding pressures Rapid curing Zero shrinkage grades available Good for moulding large parts, intricate parts, varying section parts High rigidity Good thermal resistance to +150°C Excellent electrical properties High surface finish with injection moulding Low flammability Good for insert moulding Excellent property/cost balance
Weaknesses	Handling less convenient than granular materials Limited shelf-life Susceptible to localized impact damage

Processability. BMCs can be compression, transfer or injection moulded. The use of injection moulding has become widely adopted by virtue of the economic benefits, the superior surface finishes obtainable, and the elimination of the problems of handling the sticky doughs.

The moulding of BMC represents a distinct area of thermosetting technology,which is dealt with in Chapter 7.

Typical applications

BMCs have made great inroads into many application areas in recent years. Amongst these are:

Electrical	Meter boxes and lids, switchgear, electrical housing, relays, bases
Lighting	Automotive reflectors, street lamp housings, domestic and commercial lighting
Building	Cladding panels, flooring panels
Domestic	Microwave cookware, iron handles, sandwich toasters, cooker trims
Automotive	Heater housings, trim mouldings, electrical components, inlet manifold (using lost metal insert technology)

POLYESTER SHEET MOULDING COMPOUNDS (SMCs)

Chemical family. The resin technology used in the manufacture of sheet moulding compounds is identical to BMCs, with the proviso that resin thickening is always required, and needs several days to reach completion, resulting in an essentially tack-free handleable sheet for moulding.

The same types of fillers, catalysts, pigments, etc. are also used, but in different proportions. In the case of glass fibre, significantly larger amounts and longer fibre lengths are incorporated in comparison to BMC to impart the significantly higher stiffness and strength in SMC products. Typically an SMC may incorporate 25–50% glass fibres, whereas a BMC would include 10–20%.

Form supplied. As their name implies, these compounds are manufactured in sheet form, rendering them particularly suitable for large area compression mouldings. An SMC is formed in a special machine (see Fig. 7.22) consisting of a series of rolls that are devised to bring the ingredients together and consolidate them into a sheet sandwiched between two layers of polyethylene film. The two sheets of polyethylene film are brought into the machine at different levels and are coated with a layer of polyester resin which already contains particulate fillers and catalyst system and thickening agent. Glass fibre is then chopped and deposited on one of these coated films and then sandwiched by the other.

The material thus produced is collected in rolls at the end of the machine. Such materials are supplied after storage of some days to complete the thickening reaction.

SMCs typically have a shelf-life of 3 months, during which some further stiffening can occur.

Colourability. A wide range of colours is available, including white and pastel shades.

Characteristic properties

Strengths	High strength and stiffness Excellent all round mechanical properties Good electrical performance Good thermal resistance to 175°C Outstanding dimensional accuracy and stability Zero shrink grades available High surface finish possible Rapid curing Good performance/cost balance Ideal for large surface area mouldings
Weaknesses	Restricted to compression type processes Limited shelf-life Elimination of surface porosity and internal voids in mouldings sometimes difficult

Processability. Essentially these materials are designed for large area compression mouldings. Although occasionally attempts are made to use SMCs in transfer and injection moulding processes, the author considers them to be generally unsuitable.

Typical applications. Application areas are very wide and diverse, covering many uses where high strength and stiffness are prime requirements.

SMCs have found considerable use in car body parts, truck cabs, fascias, bumper systems. It is used in many electrical components where BMC is not considered sufficiently strong, and in housings for business machines, building, furniture and the like.

POLYIMIDES

Chemical family. Polyimides include resins with both thermoplastic and thermosetting behaviour and contain the characteristic imide group:

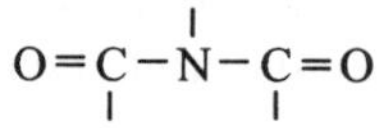

The thermosetting polyimide resins are formed by addition copolymerization of bismaleimides with amines or alkenic or alkynic monomers, with elimination of water. This is still an area of considerable development.

Form supplied. Specialized moulding materials are available containing discrete fibrous or particulate fillers for specific property balances. The main fillers include short glass fibres, graphite powder, molybdenum disulphide, PTFE and graphite fibres.

Polyimide resins are also supplied for laminating applications, and as coating and encapsulation resins for semiconductors and other electronic components. These uses fall outside the scope of this book.

Colourability. Thermosetting polyimide moulding compounds are usually filled and have a restricted colour range of dark colours and black.

Characteristic properties

Strengths	Excellent heat resistance to 270°C for long periods Non-combustible Excellent stiffness and strength Outstanding creep resistance Excellent friction and wear properties (with PTFE additions) Excellent electrical properties
Weaknesses	High cost Limited availability Processing of highest heat resistant grades Resistance to dilute alkalis and strong acids

Processability. Polyimide parts can be fabricated by techniques ranging from powder metallurgical methods to more conventional compression, transfer and injection moulding. Generally the compounds offering the highest heat resistance are the most difficult to fabricate.

Typical applications. High price has tended to dictate premium applications, although there are now indications of more competitively priced compounds becoming available.

Aerospace	Jet engine components, connector blocks, turbine engine bushings and vane bearings
Military	High reliability printed circuit boards, IC chip carriers
Automotive	Coil bobbins, gears, bearing components

POLYURETHANES

Chemical family. Polyurethane plastics encompass a wide range of solid and foamed materials. Commercialized urethanes include rigid and flexible forms, elastomers, adhesives and coatings. Urethane plastics and elastomers are available in thermoplastic and thermosetting types.

Polyurethanes, containing the characteristic urethane group –NHCOO– can be created through a variety of addition polymerization reactions involving polyisocyanates and polyols and catalysts (typically tertiary amines). The structures obtained can be complex and diverse. A commonly held model of the structure identifies the existence of 'hard' segments and 'soft' segments in the material, which contribute to the balance of rigid/elastomeric properties and general toughness exhibited by these polymers.

Form supplied. The most significant members of the thermosetting polyurethane moulding material family are those formulated for the reaction injection moulding (RIM) process dealt with in Chapter 8. These materials are true thermosetting compounds, and the reactive precursors (polyisocyanate and polyol) are supplied to the RIM process storage tanks as separate liquids in drums or larger liquid transport containers. RIM formulations may be unreinforced, or reinforced (RRIM) with short glass fibres, milled fibres or particulate fillers.

Liquid precursors for polyurethane foam manufacture are supplied in a similar manner, however, it is important to ensure water is not present in the ingredient for manufacture of solid polyurethanes, whereas water is deliberately included in foam formulations to allow release of CO_2 during the reaction.

Colourability. Polyurethane RIM moulding can be produced in a variety of colours, since the basic cured resin is (usually) a clear light brown.

Characteristic properties (polyurethane RIM only)

Strengths	Exceptional toughness Excellent surface quality Wide range of properties possible Outstanding design flexibility

	Low pressure/low energy moulding Foaming possible to reduce density Good for insert moulding Excellent for large surface area mouldings
Weaknesses	Limitations in strength and stiffness Heat resistance to 100°C Limited resistance to strong acids, alkalis, hot water and steam Controls relating to use of isocyanates Tendency to creep and sag

Processability. The reaction injection moulding (RIM) process used for converting liquid polyurethane materials entails the rapid metering and impingement mixing of the reactants in a specially designed mix head, and injection of the mixed material into a mould to complete the cure. This process is described in detail in Chapter 8.

Typical applications. This resin and moulding system is very versatile and opens up many diverse application opportunities. Some of the main areas include:

Automotive	Bumper systems, exterior panels, spoilers, trims, fascias, arm rests, head restraints, etc.
Cycles/motor cycles	Seats/saddles, instrument panels, aerodynamic enclosures
General applications	Public transport seating/furnishings, street furniture, office furniture, kitchen furniture

SILICONES

Chemical family. Silicones or polysiloxanes are semi-organic polymers with a quartz-like structure. They are characterized by the presence of an inorganic skeleton consisting of adjacent silicon and oxygen atoms –Si–O–Si–O– which imparts excellent thermal stability and chemical inertness. They also contain Si–C links by which various organic groups can be attached to the skeleton, thus imparting particular characteristics, including water repellency and dielectric properties.

The basic process of polymerization uses methylchlorosilanes as the main precursor, which are involved in a condensation reaction with elimination of water in the presence of a copper catalyst.

Form supplied. Silicones can be supplied as laminating resins, moulding compounds, varnishes, protective coating resins, casting resins, fluids and adhesives.

A wide variety of moulding compounds are available which can yield products ranging from elastomers through to hard, rigid parts. Such compounds may be supplied as one- or two-component elastomers, in a paste-like form, or as filled and unfilled gums, or as fully compounded granular moulding materials suitable for production of rigid components. Such compounds typically incorporate glass and mineral fillers.

Colourability. The resins are essentially clear and water white, thus allowing unlimited colour possibilities.

Characteristic properties

Strengths	Outstanding heat resistance to +300°C Good electrical properties over a wide temperature range Excellent durability in harsh environments Wide range of elastomeric/rigid properties possible Excellent chemical resistance Chemically inert and non-corrosive High purity grades available Flame retardant
Weaknesses	Post-cure stoving sometimes needed to improve properties Relatively expensive

Processability. Elastomeric and potting grades can be processed by a variety of methods including hand, dispenser systems, compression, transfer or injection depending on the application and consistency/viscosity of the compound.

Granular moulding compounds are suitable for compression, transfer and injection moulding. It is sometimes necessary to subject the moulded products to a post-cure stoving operation to develop full mechanical properties.

Typical applications

Elastomers	Automotive ignition harnesses, engine gaskets and sealants, hoses, ductings, rubber boots in hot areas Electronic device encapsulants and sealants Wide variety of military and aerospace components
Rigid polymers	Semiconductor encapsulation, high performance small insulators of various types

VINYL ESTERS

Chemical family. Vinyl esters have epoxy groups in the main polymer backbone which are linked by ester groups to vinyl unsaturated groups at the end of each molecule.

Vinyl group–ester group–epoxy backbone–ester group–vinyl group cross-linking takes place through the vinyl groups by an addition reaction in a similar manner to unsaturated polyesters.

While the low molecular weight results in low resin viscosity, thickening with magnesium oxide and calcium oxide is also possible.

These are relatively new materials and therefore performance and application details are still limited.

Form. Vinyl ester liquid resins can be used as a high performance replacement for unsaturated polyesters in the majority of the 'wet' moulding and fabrication processes used for making FRP products. They are also suitable for manufacture of BMCs and SMCs in the normal manner.

Characteristic properties. In comparison to polyester BMCs and SMCs, vinyl ester materials have the following attributes:

Strengths	Superior static and dynamic physical properties
	Improved glass wet out and fibre adhesion
	Improved thermal resistance to 200°C
	Improved toughness of products
Weaknesses	Relatively high cost

Mouldability. As polyester BMCs and SMCs. Because of the relative newness of these materials there is currently no published information on injection moulding.

Typical applications. First generation application studies have been focused substantially in automotive structural parts such as manifolds, valve covers, oil sumps, bumper beams, suspensions, road wheels, door structures and floor panels.

While it is not anticipated that all of these components will be commercialized, it does give some indication for potential future opportunities.

3.3 Typical properties

Table 3.2 gives typical physical, mechanical and chemical resistance values for representative compounds within the families discussed in the previous section. As such, the values can only be taken as a general guide and comparison. For specific requirements and applications it is necessary to consult manufacturers literature and data sheets.

In addition, there are now a growing number of computerized materials databanks now becoming available to assist the materials specifier and end user. Databanks such as PLASCAMS (from RAPRA Limited, UK), POLYMAT (DKI, Darmstadt Germany), PLASPEC (Plastics Technology Council, USA), POLYDATA (Polydata, Dublin, Ireland) and VKI Databank (Zurich, Switzerland) are repesentative of such generalized packages. It is also now common for individual materials manufacturers to provide materials data on their own products in computerized format.

Table 3.3 also gives general information and suggestions on processing behaviour and moulding conditions for the main families of compounds. Obviously, with the wide range of formulations available and varying difficulty of moulding of specific components, these recommendations can only be taken as a preliminary indication for a required processing operation. Again the potential user would be well advised to consult directly with the technical specialists of the materials manufacturer.

Table 3.2 Guide to thermosetting moulding material properties, part 1

Family	Allylics				Aminos						Epoxy		Polyimide
Compound	Diallyl phthalate		Diallyl isophthalate		Melamine			Urea		Melamine phenolic	Gen. purpose	High strength	Moulding grade
Property / Filler	Organic fibre	Short glass	Organic fibre	Short glass	Alpha-cellulose	Rag	Glass	Alpha-cellulose	Wood flour	Cellulose mineral	Glass & mineral	Glass	Glass
Density (g/cm^3)	1.3–1.6	1.6–1.8	1.26	1.65–1.75	1.5	1.5	1.8–2.0	1.5	1.5	1.55–1.74	1.85–1.9	1.85–1.9	1.8–2.0
Water absorption (%) (24 h/23°C)	0.1–0.3	< 0.2	0.2	< 0.2	0.3–0.6	1.5	0.1–0.3	0.6–1.0	0.6–1.0	0.3–0.5	0.05–0.1	0.05–0.1	0.2
Tensile strength (MPa)	30–50	55	30–50	55	48–90	0.3–0.6	40	35–50	35–70	30–60	70–85	115–300	140
Elongation (%)	—	—	—	—	—	—	—	—	—	—	—	—	—
Flexural strength (MPa)	50–65	90–105	50–70	90–105	80–100	80–100	90–160	80–120	70–110	50–90	105–130	260–400	250
Flexural modulus (GPa)	3–4	6–8	3–4	7–9	9.5–13	9.5	16	8–10	9–11	10–13	10	20	15–20
Unnotched impact (kJ/m^2)	2–5	> 6	> 6	> 6	> 7	> 7	> 7	> 6.5	> 6.5	> 7	> 10	10–40	15–40
Notched impact (kJ/m^2)	0.5–1.5	0.5–1.5	0.5–1.5	0.5–1.5	> 1.5	> 1.5	> 1.5	> 1.5	> 1.5	> 1.5	2–3	> 5	> 5
Deflection temperature (°C) (1.8 N/mm^2)	200	200	210	220–240	> 150	> 150	> 150	> 110	> 110	> 150	150–250	200–260	>315
Max. operating temp (°C) (short term)	180	200	190	210	100	120	150	75	75	130	120–200	150–220	270
Thermal conductivity (W/m K)	0.6–1.0	1.0	0.6–1.0	1.0	0.7–1.0	0.9	1.1	0.7–1.0	0.6–0.9	0.9	0.7–0.9	0.8–1.2	9
Coeff. of thermal expansion ($10^{-5} K^{-1}$)	2.2	1.2–1.6	1.2–1.6	1.2–1.6	2.5	1.7	2	2–5	3–6	3	1.7–3.0	1–3	1.2
Chemical resistance													
Acids weak	R					R		LR		LR	R		R
strong	LR					LR		A		A	R		LR
Alkalis weak	R					R		R		LR	R		LR
strong	LR					R		R		LR	R		LR
Aliphatic hydrocarbons	R					R		LR		LR	R		R
Aromatic hydrocarbons	R					R		R		R	R		R
Ketones	R					R		R		R	R		R
Alcohols	R					R		R		R	R		R
Halogenated hydrocarbons	R					R		R		R	R		LR

R = resistant, LR = limited resistance, A = attacked, D = dissolved.

Notes: 1. Manufacturer's specific literature should be consulted to obtained precise recommendations for specific material grades.

2. It is sometimes possible to improve the mechanical properties and thermal resistance of thermosetting mouldings by subjecting them to a post-moulding stoving operation. General recommendations for suitable stoving conditions are given in Table 3.3.

Table 3.2 Guide to thermosetting moulding material properties, part 2

Family	Phenolics			Polyesters				Polyurethanes			Silicones		Vinyl ester
Compound	General purpose	Electrical	High strength	BMC	SMC	TMC/CIC	Granular	PU RIM (H. modulus)	RRIM (Amine)	RRIM (Polyures)			SMC
Property / Filler	Wood flour	Mica	Mineral & glass	Mineral & glass	Mineral & glass	Mineral & glass	Mineral & glass	None	Milled glass	Milled glass	Glass	Fused silica	Mineral & glass
Density (g/cm^3)	1.35–1.45	1.8–1.9	2.0–2.1	1.70–2.1	1.80–2.0	1.8–2.2	1.75–2.1	1.0	1.2	1.2	1.85–1.9	1.85–1.9	1.8–1.9
Water absorption (%) (24 h/23°C)	0.2–0.4	0.05	0.15	0.15	0.2	0.2	<0.2	0.8	0.8	0.8	0.16	0.08	0.15
Tensile strength (MPa)	40–65	39–50	75–105	20–70	40–95	45–95	30–45	25–35	20–30	20–35	45	36	90–150
Elongation (%)	—	—	—	—	—	—	—	50–80	100–220	100–220	—	—	1.5
Flexural strength (MPa)	70–90	65–90	80–150	60–140	80–180	80–160	40–70				95	50	150–350
Flexural modulus (GPa)	9	13–16	12–18	7–15	8–13	8–12	6–10	0.6–0.9	0.7–1.5	1.2–1.7	17	11	12–18
Unnotched impact (kJ/m^2)	5–8	5–8	8–12	35–110	45–150	40–100	12–25	No fail	No fail	No fail	12–25	>6	50–150
Notched impact (kJ/m^2)	1.4–2.0	1.5–2.2	3–5	15–40	30–90	15–50	9–12	No fail	No fail	No fail	6–12	0.5–1.5	35–100
Deflection temperature (°C) (1.8 N/mm^2)	150–200	155	182	>200	>200	>200	>200	80	90–125	95–130	480	450	>200
Max. operating temp (°C) (short term)	110	140	150	160	175	180	160–220	60–100	80–110	85–120	320	320	>180
Thermal conductivity (W/m K)	0.8	1.6	1.3	0.7	0.5–0.7	0.5–0.7	0.6–1.0	0.2	0.2	0.2	2.0–2.5	2.0–2.5	0.2–0.3
Coeff. of thermal expansion ($10^{-5} K^{-1}$)	3–5	1.5–3.0	1.2–2.5	1–3	1–3	1–3	2–3	10–20	5–15	4–15	3.5	3	3–4
Chemical resistance													
Acids weak		R			LR				A		R		R
Acids strong		A (oxidising)			A				LR		LR		LR
Alkalis weak		LR			A				A		R		R
Alkalis strong		LR			A				LR		LR		LR
Aliphatic hydrocarbons		LR			R				R		A		R
Aromatic hydrocarbons		R			LR				R		D		R
Ketones		R			A				R		R		R
Alcohols		R			A				R		R		R
Halogenated hydrocarbons		R			R				R		R		R

R = resistant, LR = limited resistance, A = attacked, D = dissolved.

Table 3.3 Suggested processing conditions for the main families of thermosetting moulding compounds, part 1

Family	Allylics		Aminos			Epoxy	Polyimide
Compound	DAP	DIAP	Melamine	Urea	Melamine phenolic	Granular	Moulding grade
Compression							
Preheat temp (°C)	60–80	80–100	80–120	80–120	80–120	50–80	
Mould temp (°C)	150–170	150–170	160–180	160–180	160–180	150–180	200–250
Pressure (bar)	50–300	50–300	200–400	200–400	200–400	50–350	100–300
Cycle time (1–2.5 mm section) (s)	40–150	40–150	40–280	40–280	40–280	50–250	
Shrinkage from mould (%)	0.2–0.8	0.2–0.6	0.2–0.6	0.4–0.8	0.4–0.6	0.4–0.8	0.1–0.3
Post cure (if required). (temp/time)	150–190 °C/2–16 h			80–110 °C/16 h		175 °C/8 h	
Transfer							
Mould temp (°C)	150–180	150–180	140–160	140–160	140–160	150–180	200–250
Transfer pressure (bar)	350–600	350–600	500–1200	500–1200	500–1200	300–1200	400–700
Cycle time (1–2.5 mm section) (s)	35–150	35–150	35–250	35–250	35–250	40–200	
Shrinkage from mould (%)	0.2–0.8	0.2–0.4	0.3–0.8	0.4–1.0	0.5–0.8	0.4–0.8	0.1–0.3
Injection							
Cylinder temp (°C)	60–100	60–100	70–90	60–80	70–90	50–80	90–150
Nozzle temp (°C)	85–110	85–110	90–100	80–95	90–100	70–95	150–190
Melt temp (°C)	100–120	100–120	100–110	100–110	100–115	80–110	
Mould temp (°C)	160–180	160–180	155–170	145–155	155–180	160–200	200–250
Injection pressure (bar)	700–2000	700–2000	850–2500	850–2500	800–2500	300–2500	600–2500
Back pressure (bar)	5–15	5–15	5–20	5–20	5–20	5–20	5–25
Cycle time (1–2.5 mm section) (s)	30–120	30–120	20–90	20–90	25–90	25–100	
Shrinkage from mould (%)	0.2–1.0	0.2–0.8	0.3–0.8	0.5–1.0	0.8–1.5	0.3–0.9	0.1–0.3

Notes: 1. Manufacturer's specific literature should be consulted to assist in optimizing moulding conditions for a specific grade.
2. Many mouldings produced in thermosetting compounds may be subject to some post-moulding shrinkage in addition to the values quoted in these tables. In some cases it may be advantageous to introduce a post-cure stoving operation to overcome this problem, and improve dimensional stability of the component in service. This process also can be beneficial in achieving higher levels of cure, relieving of moulding stresses, and improving mechanical and thermal performance. A general indication of stoving conditions is given in the tables.

Table 3.3 Suggested processing conditions for the main families of thermosetting moulding compounds, part 2

Family	Phenolics			Polyesters			Polyurethanes	Silicones	Vinyl Ester
Compound	General purpose	Electrical	High strength	BMC	SMC	Granular	RIM/RRIM	Granular	BMC/SMC
Compression									
Preheat temp (°C)	90–110	90–110	90–110	—	—	80–100	—	50–80	—
Mould temp (°C)	160–180	160–180	160–180	130–160	135–155	150–170	—	150–170	140–160
Pressure (bar)	140–300	150–320	150–320	30–150	60–200	50–300	—	50–300	60–150
Cycle time (1–2.5 mm section) (s)	45–170	45–170	45–170	40–90	40–90	35–100	—	40–180	45–100
Shrinkage from mould (%)	0.4–0.8	0.2–0.6	0.1–0.5	0–0.4	0–0.2	0.2–0.8	—	0.1–0.6	0–0.2
Post cure (if required) (temp/time)		150–190 °C/16 h			120–170 °C/8 h		30 min/120 °C	250 °C/8 h	175°C/8 h
Transfer									
Mould temp (°C)	160–180	160–180	160–180	150–170	150–170	150–180	—	150–170	140–170
Transfer pressure (bar)	450–700	450–700	600–900	100–500	150–600	450–850	—	300–700	100–500
Cycle time (1–2.5 mm section)	35–115	35–115	35–115	40–85	40–85	30–90	—	40–180	45–85
Shrinkage from mould (%)	0.5–1.0	0.3–0.7	0.2–0.6	0–0.4	0–0.2	0.2–1.0	—	0.1–0.6	0–0.2
Injection									
Cylinder temp (°C)	60–85	60–85	60–85	20–60	—	50–80	—	50–80	20–50
Nozzle temp (°C)	80–95	80–95	80–95	20–40	—	80–95	—	70–90	20–50
Melt temp (°C)	110–125	110–125	110–125	30–80	—	100–120	35–45	80–110	30–80
Mould temp (°C)	160–185	160–185	160–185	140–170	—	160–180	60–90	160–200	140–170
Injection pressure (bar)	600–2500	600–2500	600–2500	250–1500	—	300–1200	10–50	300–2500	250–1500
Back pressure (bar)	5–20	5–20	5–20	5–10	—	5–15	—	5–15	
Cycle time (1–2.5 mm section) (s)	20–80	20–80	25–90	35–90	—	30–80	45–150	30–100	0–0.2
Shrinkage from mould (%)	0.6–1.0	0.3–0.7	0.2–0.6	0–0.4	—	0.2–1.0		0.1–0.6	

CHAPTER 4

Additives, fillers and reinforcements

4.1 Introduction

As the reader will have already appreciated, thermosetting resins usually require to be mixed with a variety of additional materials to arrive at a formulation which can be considered as an exploitable moulding compound.

The range of additives from which to choose is obviously very great, and the number of theoretical combinations limitless. It is therefore of little use simply to list the possibilities that are open to the formulator, since such an approach would only serve to confuse the reader. An attempt is therefore made to provide some structure to the subject by starting with a discusion of the main functions required of additives in thermosetting compounds in the first part of the chapter, before going on to discuss typical ingredients that fulfill these requirements in the second part.

For the purposes of this work, catalysts, including reaction promoters (or initiators) are not treated in this chapter, since they are considered to be an integral part of the resin system. As such, they are mentioned elsewhere in the book in direct connection with the specific resins and cure mechanisms to which they are linked.

The role of fillers and reinforcing fibres is a central issue, since they play the major part within the compound in facilitating the creation of the moulded composite. It is the role of the polymer matrix to bind reinforcing fibres and particulate fillers together into a composite structure, and thereby enable the transfer (and sharing) of applied load between all fibres and particles in the composite, and prevent crack propagation. The fillers can therefore contribute to significant improvements in composite physical properties and also to cost reductions.

Before going on to discuss the subject in more detail it is advisable to pause to consider the main categories of additive in relation to the principle functions they serve within the compound.

The classification and examples given in Table 4.1 are of a general nature, and therefore applicable to thermosetting and thermoplastics applications. By now, it will be appreciated that the specific knowledge and experience required to select and use these additives alone or in combination, within a formulation is not acquired easily. The role of the compound formulator can be seen to be a specialist function, based on a skill built up over some years in the appropriate field. This discussion, however, will serve to make the non-

Table 4.1 Main categories of additives for plastics

Main categories	Specific types
1 Additives which modify bulk mechanical properties	a. Plasticizers or flexibilizers
	b. Reinforcing fillers
	c. Toughening agents
2. Additives used to reduce formulation costs	a. Particulate fillers
	b. Diluents and extenders
3. Optical property modifiers	a. Pigments and dyes
	b. Nucleating agents
4. Anti-ageing agents	a. Anti-oxidants
	b. UV stabilizers
	c. Fungicides
5. Surface property modifiers	a. Antistatic agents
	b. Slip agents
	c. Anti-wear additives
	d. Anti-block additives
	e. Adhesion promoters
6. Additives which assist processing	a. Lubricants — internal
	— external
	b. Processing aids and flow promoters
	c. Thixotropic agents
	d. Processing stablizers
7. Others	a. Blowing agents
	b. Flame retardants

specialist aware of some of the more important choices and more common ingredients that are likely to be encountered within the industry.

4.2 Fillers – general aspects

The classification of materials (usually loosely called fillers) which can modify composite physical, mechanical, thermal or electrical properties is not straightforward. One distinction can be by mode of occurence, for example, organic, inorganic or synthetic. In Table 4.2 we can see a selection of naturally occurring fillers together with comments on their main properties.

An alternative classification is by relationship to filler role, for example reinforcing (normal with respect to mechanical properties), heat resistant, conductive or non-conductive and so on. A third is by geometry. for example spheres, cuboidal, angular, platelets or fibrous.

In general terms the geometry of the filler determines the reinforcing potential, with fibrous fillers (e.g. asbestos, glass, wollastonite and some organic fibres such as flax and jute) being very effective. Other organic materials such as alpha cellulose, cotton flock, paper, fabrics, sawdust and woodflour have fibrous structures and also provide a good level of reinforcement to the composite. At the bottom of the scale we find some of the cheaper mineral fillers whose primary purpose is to reduce compound cost.

Table 4.2 Naturally occurring fillers for plastics

Material	Properties
Alumina trihydrate	Fire retardant, decomposing at 170–200 °C
Barytes (barium sulphate)	Specific gravity of 4.5 makes this a dense filler; used for example, for acoustic insulation
Hollow aluminium silicate (perlite, fly ash and spheres)	Glassy materials used in low density applications, including floatation equipment
Organic materials (cellulose, woodflour, starch)	Used to reinforce thermosets such as ureas and phenolics
Calcium carbonate (whiting, marble, limestone)	Occurs naturally as calcite, extensively used in almost all polymers to reduce cost and improve properties
China clay (hydrated aluminium silicate)	Popular general purpose filler: thin hexagonal crystals provide reinforcing effect in some polymers; occurs as kaolin
Talc (hydrated magnesium silicate)	Probably the highest tonnage filler; long laminar flakes offer good reinforcement in plastics; often contains much impurity
Mica (different types of aluminosilicates)	Flat plate structure gives good reinforcement
Calcined clay	Heated kaolin forms dehydroxylated silicate; hard and thermally stable; retains plate structure

Particulate fillers can also provide useful reinforcement where the particle size and geometry is favourable. Common examples are mica, talc, calcium carbonate (calcite), china clays, milled glass and glass spheres. Enhancement of mechanical properties is usually lower than with fibrous reinforcements, although more comparable for impact behaviour of the composite. They can, however, also often bring additional property benefits in the areas of electrical and thermal properties, and are usually much cheaper than fibrous fillers. All of these materials will significantly reduce the moulding shrinkage of the composite (see Fig. 4.1) and also the coefficient of thermal expansion, thus leading to more dimensionally predictable and stable products. These fillers will also have a significant influence on the overall density of the moulded composite.

Developing the argument of filler classification by geometry type we can arrive at a typical listing as shown in Table 4.3.

A common method of comparison of filler geometry is by aspect ratio, which is defined as:

$$\text{aspect ratio} = \frac{\text{length}}{\text{diameter}} \qquad [4.1]$$

A microscopic examination of some typical fillers shown in Fig. 4.2 highlights the significantly different structures that can be found. The different forms of glass reinforcement are dealt with in a later section of this chapter. Most mineral fillers are crystalline solids that break along preferred crystal boundaries, giving rise to filler particles of defined shape (cubic,

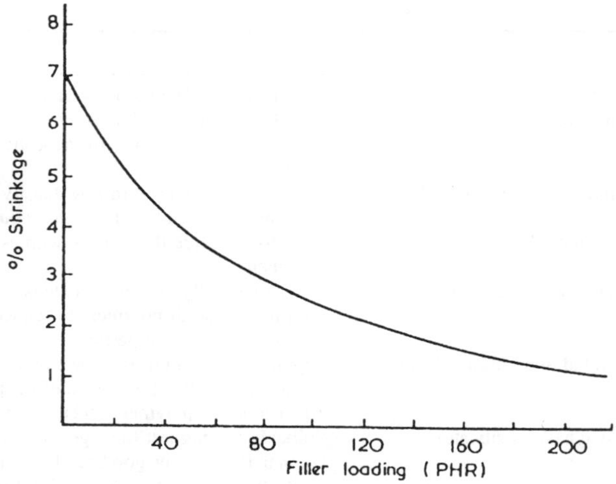

Figure 4.1 Effect of filler loading on the shrinkage of an unreinforced polyester composite (filler in this case is an uncoated calcium carbonate).

Table 4.3 Some typical fillers classified by geometric type

Spherical	Cuboidal or Angular	Platelets	Fibrous
Glass beads	Calcite	Clay	Glass fibre
Glass or ceramic bubbles (microspheres)	Dolomite	Mica	Wollastonite
		Quartz	Asbestos
			Calcium carbonate fibre

rhombohedral, flake, fibre, etc.). Although these basic shapes are difficult to observe in the mineral filler, because of damage and interferences during the forming stages, they still have a profound effect on the performance of the polymer composite.

Before proceeding further it is important to understand the manner in which fillers are defined, and several terminologies associated with fillers.

Particle size. The size of a filler can only be accurately defined by its complete particle size distribution, which is normally plotted as a distribution curve. Care has to be taken in not classifying the material by its mean particle size alone, since two fillers with the same mean particle size can vary significantly in distribution (Figs 4.3 and 4.4)

Particle size distribution. The distribution can be varied to some degree by changes in processing conditions, or by blending of fine and coarse fractions. The traditional method to measure particle size distribution is with a series of graded sieves. Today the most common approach is to use a density technique involving a hydrometer, assisted by automatic instruments.

Top cut. Refers to the maximum size particle in the filler.

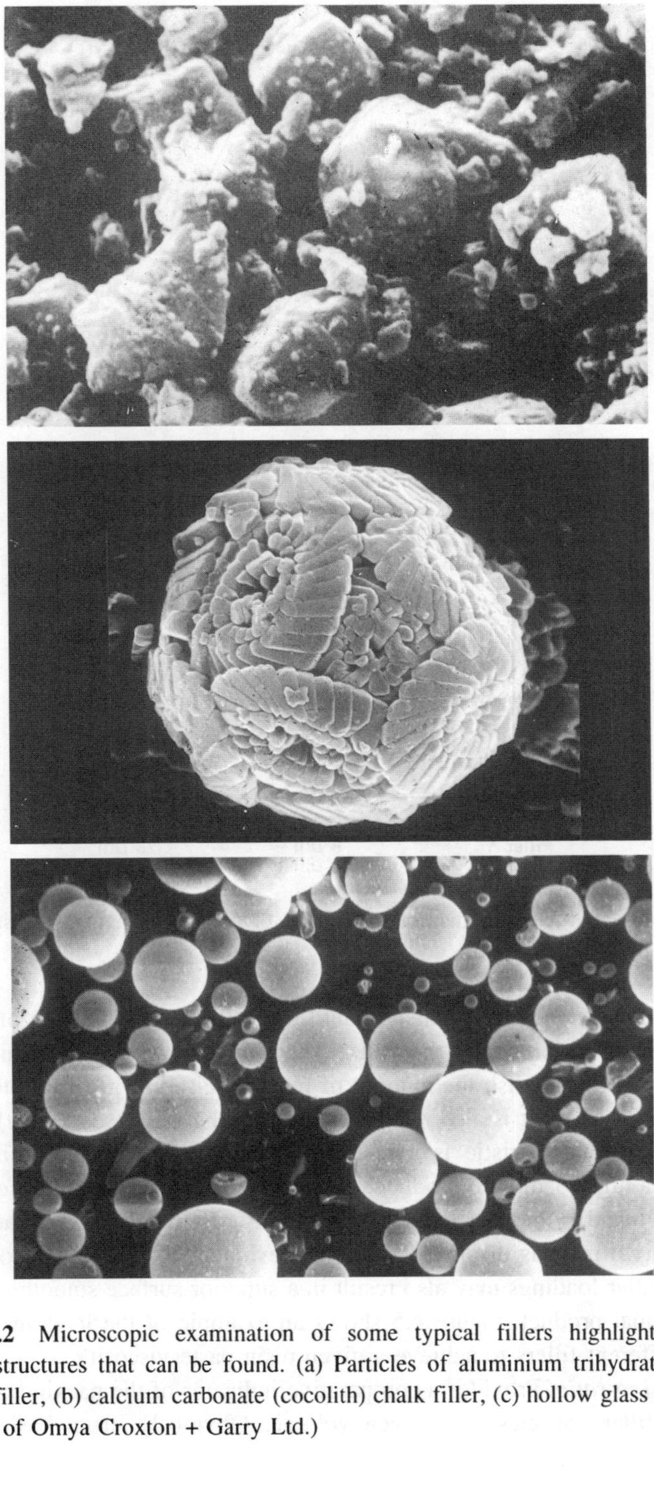

Figure 4.2 Microscopic examination of some typical fillers highlighting the different structures that can be found. (a) Particles of aluminium trihydrate flame retardant filler, (b) calcium carbonate (cocolith) chalk filler, (c) hollow glass spheres. (Courtesy of Omya Croxton + Garry Ltd.)

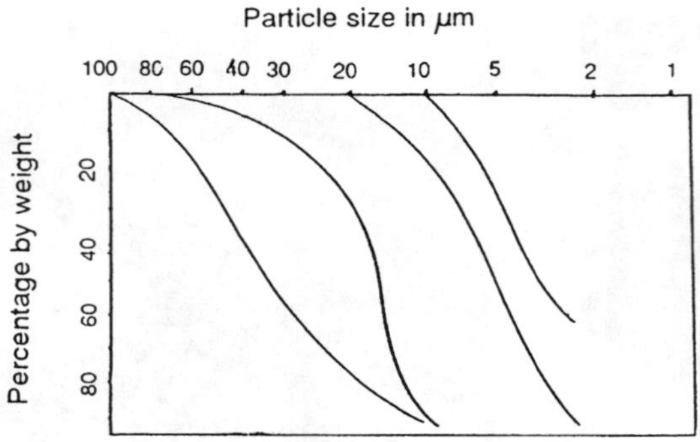

Figure 4.3 Particle size distribution curves of fillers from Orgon Site.

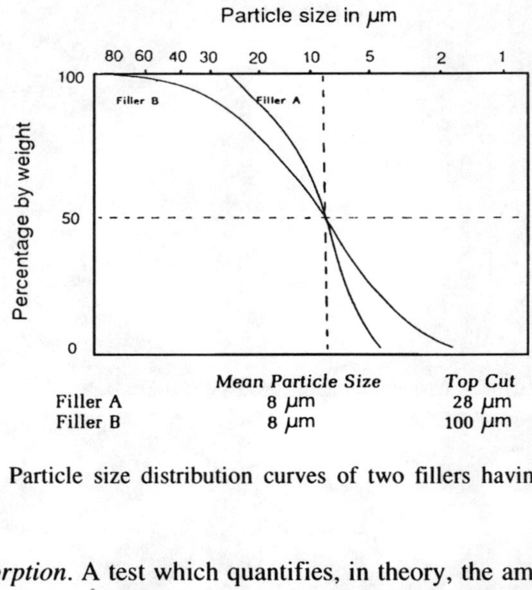

	Mean Particle Size	*Top Cut*
Filler A	8 μm	28 μm
Filler B	8 μm	100 μm

Figure 4.4 Particle size distribution curves of two fillers having the same mean particle size.

Oil absorption. A test which quantifies, in theory, the amount of polymer that will be absorbed by the filler. The simplest way of determining this property is to establish the amount of resin required to just saturate a given quantity of filler. It is also related to the packing ratio (see below). The oil absorption characteristic is sometimes called 'resin demand'. Fillers that require a high loading to produce a given viscosity (i.e. low oil absorption or resin demand) yield the greatest compound cost reductions, because more (expensive) resin is displaced in the achievement of a given viscosity. The higher filler loadings may also result in a superior surface smoothness rating in the final product. Figure 4.5 shows an example of the loadings required with different fillers to achieve a given resin paste viscosity.

Packing ratio. Provides a comparative indicator of the packing efficiency of the filler particles in a given volume. Obviously, from basic particle

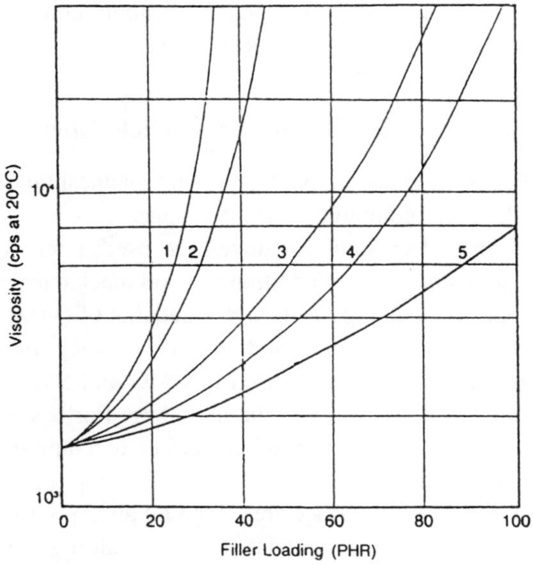

Figure 4.5 Loading of different fillers to develop a given paste viscosity. (1) Talc, (2) calcium carbonate, (3) clay, (4) whiting, (5) treated calcium carbonate. (PHR = parts per 100 parts resin.)

packing theory, irregularly shaped particles will pack inefficiently and have more void spaces between the particles.

Filler density. The Einstein equation for the rheology of a filled system can be expressed as:

$$\eta = \eta 0(1 + \kappa_f U_f) \qquad [4.2]$$

where η is the viscosity of the filled resin, η_0 is the viscosity of the neat resin, κ_f is the filler aspect ratio, and μ_f is the volume fraction of resin displaced, shows that the viscosity of the filled system increases with the volume fraction of the filler. It follows that a greater weight of high density filler is required to develop the same viscosity obtained with a low density filler.

4.3 Glass spheres and bubbles

Glass spheres or beads are most advantageously used as a filler in situations where the formulator wishes to improve the flow behaviour of the compound, particularly in thin section mouldings or mouldings having very long flow length elements. In addition, glass spheres have the benefit of reducing differential shrinkage and warping tendency in mouldings. Use of glass beads can therefore be a useful additive in cases where high filler loadings would otherwise result in very stiff moulding compounds.

Applications are most common in granular phenolic, epoxy, and granular polyester formulations as well as many RIM systems.

4.4 Cuboidal or angular particle fillers

Calcite (calcium carbonate) and dolomite (calcium magnesium carbonate) are listed as two of the more commonly used examples.

Rhombohedral shaped crystalline calcites are used extensively in SMC and BMC formulations because, in addition to good mechanical properties, it is also usual to require mouldings with high standards of surface finish. The fineness of these fillers is excellent and in addition they aid flow during moulding. Dolomites are also widely used in BMC and SMC formulations because they confer good whiteness to the moulding. For best surface finish 'automotive' type SMCs, where an excellent surface is required, a crystalline calcite with a mean particle size of 3 μm and a top cut of 10 μm is recommended. In BMCs, high surface area (high oil absorption) fillers can be used to control the flow of the material during moulding. Clays, talcs or precipitated calcium carbonates are often used to achieve this result. Such products 'mop up' resin and increase viscosity.

Calcite fillers and dolomites are also commonly used in phenolic and epoxy moulding compounds and other similar applications.

4.5 Platelets

Fillers with platelet geometric structure, such as clays, mica and quartz, provide significant reinforcement while also giving good flow characteristics during moulding, resulting from the easy movement possible between platelets.

Mica is particularly useful since, in addition to the above mentioned properties, it is particularly good in enhancing the electrical insulation characteristics of the moulding compound, as well as providing a high level of resistance to significant temperature excursions. This is often exploited in phenolic, epoxy and granular polyester moulding compounds.

4.6 Fibrous fillers

A very wide range of fibrous materials is available for specialist reinforcement applications. Discussion of such a range is beyond the scope of this book and can be followed elsewhere.[31] Glass fibres are undoubtedly the most extensively used fibrous reinforcement for polymer matrices.[32] They are formed by drawing molten glass through small orifices (known as 'tips') in a special alloy trough (bushing). Many such filaments, each finer than a human hair, are drawn at the same time and brought together in a gathering

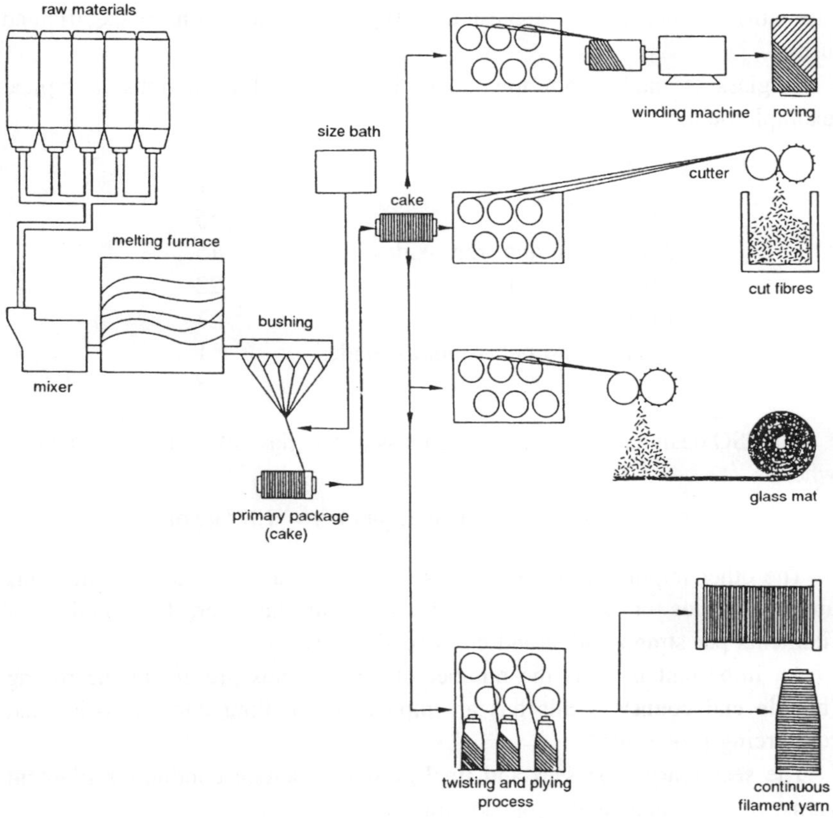

Figure 4.6 Flow diagram of the glass fibre manufacturing process. (Source PPG Fiberglass BV.)

shoe before winding into a primary package. It is during this winding stage that a most important feature, the size coating, is applied from an aqueous solution or dispersion. The primary package is then dried before subsequent processing into the final product (Fig. 4.6).

The purpose of the size coating on the fibres is to:

- adhere the fibres together and protect the glass from abrasion
- lubricate the fibres during subsequent processing
- provide good chemical bonding between the fibres and the polymer matrix
- impart antistatic properties to the fibres.

Size coatings are made from complex mixtures of polymers, lubricants and coupling agents.

A number of different glass types can be used, 'E' (or electrical) grade being by far the most common. Others include 'C' (chemical) and 'S' (high strength). As their names suggest, each type is designed to have specific end uses.

E glass is made from low alkali borosilicate formulations, a typical example being:

	%
Silica	55
Aluminium and iron oxides	15
Lime and magnesia	22
Boric oxide	7
Sodium and potassium as oxides	1

The ISO designation for the linear mass of the glass fibre roving is the tex, where:

$$1\,\text{tex} = 1\,\text{g}\,\text{km}^{-1}\ \text{fibre length (or 1 tex} = 10^{-6}\,\text{kg}\,\text{m}^{-1})$$

The other important characteristics of a roving are whether the filaments are continuous or discontinuous, the filament diameter, the number of filaments per strand and strand cross sectional shape.

An important issue is the number of bundle ends present in the roving (bundle end count) as this has an impact on handling characteristics and reinforcing properties.

The secondary processes and final products include continuous filament yarns, rovings, mats and chopped strands.

4.6.1 Continuous filament yarns

These are made by twisting and plying a number of strands together. The total number and diameter of the filaments affect the strength and flexibility of the yarn while the twisting or plying affect its smoothness and handling. Glass fibre fabrics are woven or knitted on standard textile machinery from the yarns and the resultant cloth can be used as the basis for laminating (printed circuit boards) or for coating with elastomers or thermoplastics to give good heat or weather resistant fabrics.

4.6.2 Rovings

A roving consists of an assembly of glass strands wound together without any mechanical twist to form a tubular cylindrical package. This can be used in a variety of applications.

Filament winding. In this process, the roving is wound onto a shaped mandrel after passing the tow through a resin impregnating bath. The resulting pipe or cylinder is cured and the mandrel removed. Very strong, high glass content laminates can be formed in this way.

Spray-up roving. In this application, the roving is chopped at high speed through a gun which also sprays catalysed resin. The glass and resin meet in mid-flight and form a web on the mould surface. This is consolidated by rolling and subsequently cures to give the finished article.

Panel roving. Here a large number of rovings are chopped onto a moving bed of resin. This resin is then cured to give a translucent panel which can either have a flat or a corrugated profile. These panels are used for industrial or horticultural roofing.

SMC roving. Sheet moulding compound is made by chopping roving onto a moving bed of a polyester/filler paste. This paste thickens to a non-tacky, 'leathery' consistency during the course of a few days. This process is ideal for high production rates, and is covered more fully in Chapter 7.

4.6.3 Mats

There are a number of different types of reinforcing mats, but they fall into two main groups.

Continuous filament mat. As its name implies, this mat is made by laying down a web of continuous fibres and then binding them together within a soluble or non-soluble binder depending on the application. These mats lend themselves to applications where the reinforcement has to 'stay put', as in resin injection, pultrusion or compression moulding.

Chopped strand mat. This type of mat is probably the most familiar, being the sort of material used for boat building and car repairs. It is made by chopping fibres to form a web and binding it together with either a powder or an emulsion system which imparts the right handling and moulding characteristics for the end application.

4.6.4 Chopped strands

This category covers by far the widest range of products. Fibres can be chopped into many lengths from 1.5 mm up to 25 mm and can be presented in filament diameters from 9 to 14 μm and a number of strand texes. However, it is the size coating which leads to the biggest variety, because it is this coating which renders the product suitable for its end application. Consequently there are products available for many matrices.

4.7 Plasticizers and flexibilizers

Such materials come under the general heading of additives which assist processing and are therefore discussed in the following section to avoid confusion. Since these additives are sometimes used to modify the resin reaction rate, and the corresponding rate of viscosity change during the curing reaction, it is possible to use them to tune the material formulation to arrive at 'easy flow' moulding grades.

In some cases the plasticizer may provide the additional properties of increasing the flexibility and reducing the brittleness of the final product. This can be achieved by shifting the resin glass transition temperature to a lower value by decreasing intermolecular forces, thereby having an overall effect of reducing the elastic modulus and melt viscosity.

4.8 Toughening agents

With the requirement for thermosetting composites to be used in even more demanding mechanical applications, such as vehicle underbonnet components, there has been increasing activity in the development of toughening agents to enhance the general toughness and impact strength of the final product.

With many thermosetting materials, increasing cross-linking density is accompanied by higher stiffness. In these situations it becomes more difficult for the resin to absorb energy and yield under stress, resulting in a tendency to brittleness.

Rubber toughening with liquid rubber compounds has been successfully used to control these problems, with phase separation of the rubber occurring during the resin cross-linking, to provide shear yielding sites in the composite. These rubbers can, however, reduce the resin's glass transition temperature and modulus.

An alternative approach which has been developed is to use linear high molecular weight thermoplastic polymers to provide the toughening function. Considerable success has been reported in the toughening of phenolic and epoxy resin based moulding compounds.[33]

4.9 Diluents and extenders

As the names imply, these ingredients may be included in the resin component of the formulation as materials that can effectively dilute and extend the (more expensive) resin to serve a larger total mix quantity. Obviously, such an approach can result in some reductions in the overall performance of the composite, depending on the choice of extender and quantity used.

4.10 Colourants

Colourants fall into two categories – pigments or dyes.

Pigments are distinctively coloured compounds with a refractive index much higher than the polymer, and having a particle size larger than the wavelength of light. They then impart colour and opacity to the polymer system.

Organic pigments such as benzidine and its derivatives (for yellow–orange shades), copper phthalocyanine (blue shades), chlorinated copper phthalocyanine (green–blue shades) and airline black (black–bluish and green shades) are dyes that have been rendered insoluble by thermosetting binders. They are therefore less migratory and more thermally and chemically stable than dyes. Carbon black is an ultraviolet absorber, antioxidant and reinforcement.

Inorganic pigments fall into two main categories, the first comprising salts and oxides of metals, for example titanium dioxide (white), lead chromate (yellow–orange), chromium oxide (greens) and sodium aluminium polysulphide silicates (ultramarine blues). The second category includes metallic flakes (aluminium, copper, copper alloys) and pigments providing pearlescent effects (e.g. basic lead carbonate). These pigments are insoluble and non-migratory, and usually less bright than organic dyes. They have good thermal and chemical resistance.

4.11 Anti-ageing additives

These ingredients protect the polymer from degradation during processing and throughout its useful service life. They normally fall into four classes: processing stabilizers, antioxidants, ultraviolet stabilizers, fungicides.

There can be various causes of degradation, including heat, oxygen, ozone, light, transition metal salts, chemical attack, microbiological attack, etc. Degradative influences are commonly synergistic.

Melt (processing) stabilizers, such as quinone and butylated hydroxytoluene react with polymer active free radicals to create more stable radicals, thus reducing the rate of autocatalytic degradation of the polymer.

Antioxidants, on the other hand are substances incorporated in polymers at low concentrations to prevent oxidation. Melt stabilizers can be included in this group since they can also operate in this mode. Ultraviolet stabilizers, such as the benzophenones and the benztriazoles, preferentially absorb ultraviolet energy, and in so doing protect the polymer and thereby reduce the tendency to free radical formation.

4.12 Surface property modifiers

This group encompasses antistatic agents, slip additives, anti-wear additives and adhesion promoters.

Antistatic agents reduce or eliminate build up of electrostatic charges on finished products by rendering the surface conductive. They are therefore usually ionizable additives which migrate to the surface of the moulding and form conductive paths through the absorption of atmospheric moisture. Examples are long-chain amines, amides, sulphonic acid, sulphonates and polyhydric alcohols. Antistatic problems tend to be far more common with thermoplastics than with thermosets.

Slip additives impart reduced friction to surfaces of finished products. The additives must be incompatible with the polymer so that migration to the surface can occur. Typically they are used to reduce friction and wear in moveable parts, and to reduce operating temperatures in bearings, gears and cams. Some examples are silicone fluids, molybdenum disulphide, graphite, PTFE.

Adhesion promoters can be very diverse in constitution to serve a wide range of needs. Their use is more common in thermoplastic compositions, where the basic characteristics of the resin and absence of substantial amounts of filler can in some cases lead to polymer surfaces which provide adhesion problems.

4.13 Additives which assist processing

This group includes internal and external lubricants, processing aids and thixotropic agents.

Internal and external lubricants fulfil two complementary functions. The internal lubricant is used to improve plasticization at high temperatures and increase the fusion rate of polymer particles when rubbing against one another. External lubricants on the other hand are used to reduce interparticle friction between (a) polymer granules or powders (in the solid state), (b) polymer melt and metal surfaces of processing equipment (including moulds), and (c) to reduce friction in the finishing operations such as packaging.

It is not surprising, in view of the relatively close nature of these needs, that internal and external lubricants are chemically similar. Typical materials are high molecular weight fatty acids, alcohols, amines, amides, metal soaps in the C_{13} to C_{18} range and high molecular weight hydrocarbon waxes.

In some cases a particular lubricant will fulfil both functions (internal and external), but in general internal lubricants require to have greater compatibility with the polymer.

The primary function of processing aids is to act as flow promoters by bringing about plasticization of the polymer at high temperatures. In this respect they are similar to internal lubricants. Their general effect is to increase fusion rate of polymer particles and to improve the flow characteristics of the melt during processing. With thermosets in particular, flow promoters are used to reduce the rate of interaction of reactive groupings by exercising a dilution effect. In this way they tend to reduce the rate at which the melt viscosity increases as a result of cure onset. Typical processing aids are low molecular weight polymers or waxes.

Thixotropic additives can be used to formulate compounds having a relatively high liquid or melt viscosity in static condition, which can be reduced significantly by shearing action.

4.14 Blowing agents

Blowing agents are used to produce cellular materials. Physical blowing agents can be low boiling range liquids or gases that are soluble in the polymer matrix and which provide a blowing action as temperature is increased and/or pressure reduced during processing. Chemical blowing agents are additives which generate gas within the compound as a result of chemical decomposition of the additive. The decomposition may be caused to occur at a controlled rate over a narrow temperature range, or alternatively it may be catalysed or inhibited.

The uses and technology of foamed plastics employing blowing agents has been more developed and widely exploited with thermoplastics than with thermosets in the past. The most notable exception to this picture is in the area of polyurethane RIM and RRIM, which is covered in Chapter 8.

The creation of cellular structure in the moulded product not only reduces the density of the material, but also modifies many other properties including mechanical, thermal, acoustic and electrical performance.

4.15 Flame retardants

Flame retardance of plastics is rapidly gaining importance and there are numerous possibilities for formulating flame retardent resins. Many thermosets such as phenolics, silicones, polyester alkyd, DAP, DIAP, UF, MF and polyimides have intrinsic flame resistant properties. Other resins, such as polyesters, can be made flame retardant by incorporating antimony trioxide, halogens, phosphorus compounds, boron compounds, nitrogen compounds or alumina trihydrate.

In recent years much pressure has been applied to formulators and designers to use plastics that are not only flame retarded but also exhibit low smoke density and low smoke toxicity, should they actually burn.

When plastics burn, carbon, carbon monoxide, carbon dioxide and water vapour are developed. Hydrogen chloride and hydrogen bromide are generated from halogenous flame retardants. Antimony trioxide and halogens form antimony perchloride and antimony trichloride. Nitrogen-containing flame retardants evolve hydrogen cyanide. Flame retardants reduce the

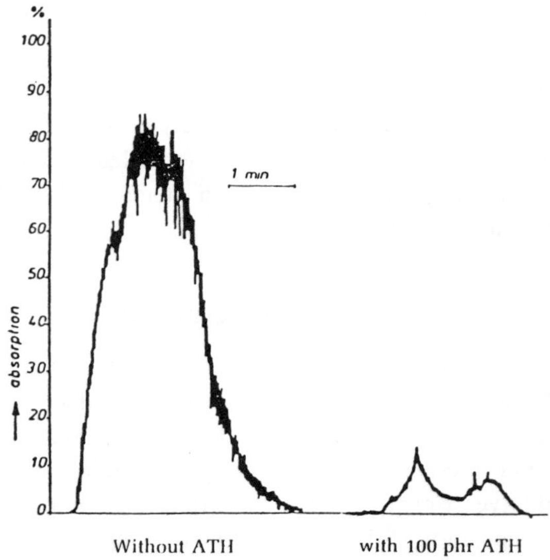

Figure 4.7 Smoke development of an unsaturated polyester resin with and without aluminium trihydrate (measured by absorption spectra).

complete decomposition into carbon dioxide and water vapour and enhance the formation of incompletely oxidized decomposition products, for example carbon and carbon monoxide. The use of flame retardants thus often increases the smoke density and emission of toxic gases.

Aluminium trihydrate (ATH) on the other hand decomposes into the physiologically harmless products aluminium oxide and water. Furthermore, it inhibits soot formation and thus reduces the smoke density.

The effect on smoke reduction can be measured by light absorption techniques, with a typical test result shown in Fig. 4.7.

4.16 Shrinkage control additives

These are additives which can be incorporated into bulk moulding compounds (BMCs) and sheet moulding compounds (SMCs) to control more accurately and/or reduce the mould shrinkage of these compounds.

The additives used are usually finely divided thermoplastic powders or synthetic elastomers. A summary of typical ingredients used for the main families of BMC/SMC compounds is given in Table 4.4.

The first category of additives in Table 4.4 are non-polar materials which are non-compatible with the resin system and added in quantities up to 5–10%. Although some shrinkage reduction is achieved with these additives, the main benefits come in the improved surface finish and pigmentability.

The other two groups of additives have been developed with main aim of

Table 4.4 Additive families developed for shrinkage control properties

Category	Examples
1. First generation Non-polar polymer additives which are non-compatible with the resin system	Polystyrene Polyethylene
2. Second generation Systems with reduced or no moulding shrinkage developed for automotive and other commercial applications	Polyvinyl acetates Polymethylmethacrylate and co-polymers Polyvinyl chloride – vinyl acetate co-polymers Styrene – butadiene co-polymers Cellulose acetate butyrate Modified polyvinyl acetates
3. Third generation Highest levels of shrinkage control accuracy of mould reproduction and surface smoothness	Dual thickening systems containing isocyanate prepolymers Saturated polyesters Saturated polyester blends with polyvinyl chloride Polyurethanes

significantly reducing and controlling moulding shrinkage. They operate with a different mechanism to the first group, being additives that are to some extent soluble in the uncured polyester resin, but subject to phase separation during the resin cross-linking reaction. With polar additives, such as polyvinyl acetate, polyvinyl chloride, acrylic polymer and others, the influence of changing polarity of the polyester resin (from polar in the uncured state to non-polar in the cured state) is an important factor in the phase separation behaviour.

Further discussion of the shrinkage control mechanism is given in subsection 7.1.1.

For the reader who wishes to go further into details of filler and additive technology, useful guidance is given in a number of reference works[34–36]

CHAPTER 5

Compression and transfer moulding

5.1 Compression moulding

Compression moulding is the oldest method of mass production of plastics components, and still retains an important role by virtue of a combination of benefits including:

- simple process — short route from raw material to finished product;
- material movement in the mould is limited, thereby promoting isotropy, and reducing tendencies towards directional orientation and retained internal stresses in the product;
- mechanical and electrical properties of the material remain high because little mastication and reinforcement breakdown occurs in the process;
- output rates can be high with multi-cavity moulds;
- capital equipment and moulds are relatively simple and less costly compared with injection moulding.

5.1.1 Process description

Fundamentally the process consists of placing a controlled amount of moulding material in powder or tablet form into a heated metal mould, closing the mould and applying a suitable clamping force while the curing reaction takes place. A schematic representation of the moulding sequence is given in Fig. 5.1.

When the moulding material is introduced into the mould the resin in the compound fluxes with the heat, and the whole mass is pressed into the shape of the mould cavity. Polymerization proceeds and after a suitable interval the mould is opened and the moulding removed. The cross-linking reaction ensures that the moulding is rigid and infusible (although still hot) at this stage.

If the moulding is to be produced from loose powder, it is possible to calculate the weight of the charge knowing the volume of the cavity, the true density of the cured material and making a small allowance (maximum 5%) to ensure filling and a minimum amount of flash (excess) material to be expelled from the cavity. Alternatively the loose powder charge can be dry pelleted in a tabletting press with a compression force of 900–1200 bar. Pellets are normally cylindrical or spherical, and of standard diameters.

Optimum placement of the charge in the mould cavity requires some

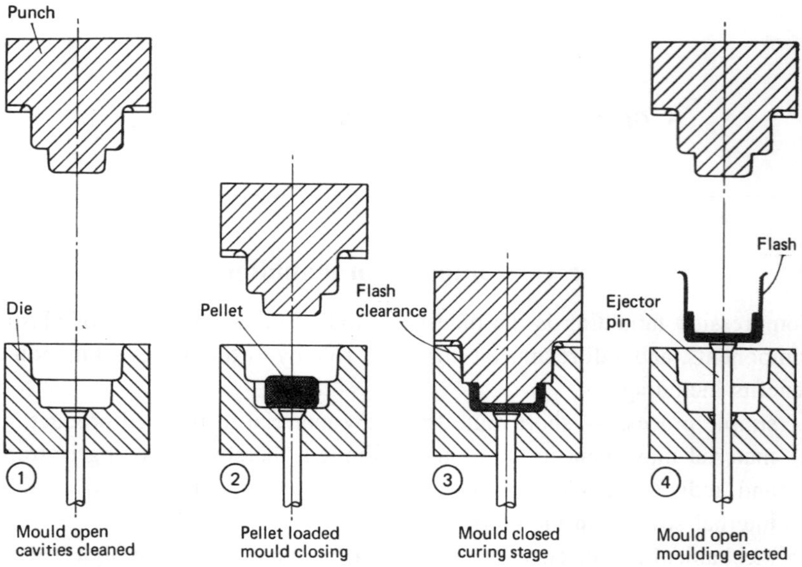

Figure 5.1 Sequence of compression moulding operations.

consideration, particularly if the component has a relatively large surface area or geometric complexities that could lead to the need for significant material movement during the fluxing phase. This is particularly important when a simple pellet is used, or powder deposition into the cavity is too localized.

For the majority of thermosetting moulding compounds, compression moulding is carried out in polished steel moulds heated to obtain a mould surface temperature of within the range 140–170°C. Heating of the moulds is normally by electricity, steam or hot water under pressure. Stainless, high carbon, high chromium mould steels are usually specified, and are heat treated to obtain an optimum balance of surface hardness, wear resistance and toughness. During fluxing the resin behaves as a viscous fluid and transmits pressure. The fillers and reinforcements however, tend to impede this effect and also the resin becomes increasingly viscous as cross-linking progresses. Moulding pressures depend on the characteristics of the compound to be moulded, plus specific effects of the component and mould design. A general guide to typical moulding pressures for the more common families is given in Table 5.1.

Set (press) moulding pressures for a given product must be calculated in relation to product geometry. The basic principle of the calculation is always the same, but different materials manufacturers will provide specific recommendations for their compounds. For example, one well-known supplier of granular phenolic moulding compounds recommends a pressure requirement for cold powder moulding in a semi-positive mould of 210 bar of landed area up to 25 mm of cavity depth. An additional 50 bar is then

Table 5.1 Typical moulding pressures

Material type	Moulding pressure (bar)
Polyester bulk moulding compounds (BMCs)	60–100
Granular polyesters and soft-flow phenolics	140–280
Amino plastics and stiffer-flow phenolics	200–400
Stiff-flow materials	420–560

required for each additional 25 mm of depth.[37] If preheated material is used, the pressure requirements may be significantly reduced by, typically, one third of these levels. The effect of mould temperature on moulding pressure and material flow time is illustrated in Fig. 5.2 for a typical phenolic compound.

The mould closing phase of the process can be more complex than is apparent from the foregoing description. Firstly, it is normal practice to restrict the press closure speed towards the end of the stroke to avoid excessive impact between the mould halves, minimizing risk of damage to the mould elements and also avoiding the possibility of 'splashing' moulding powder out of the cavity. Secondly, it is common practice to slightly relax the clamping pressure on the mould for some seconds shortly after closure and during the period in which the moulding compound has fluxed and is filling the mould. This operation is normally referred to as 'breathing' the mould, as it is allowing easy escape of trapped air and the gases produced during the curing reaction. Failure to do this can sometimes result in components with inferior mechanical and electrical properties and poor dimensional tolerances owing to internal porosity, flow lines and other internal defects relating to the trapping of excessive levels of gas and volatiles. It is advisable, if possible, to

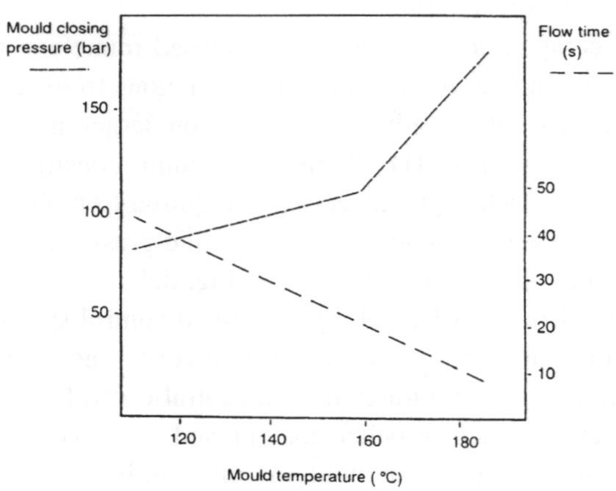

Figure 5.2 Effect of mould temperature on the mould closing pressure and material flow time for a general purpose phenolic compound.

regulate the process closing speed to maintain a constant speed of movement of the material flow front.[38]

The time required for complete polymerization of the moulding is governed by the reactivity of the compound and the geometry of the component. Maximum section thickness is obviously the critical factor. As mentioned in Chapter 1, the majority of thermosetting resins shrink during the cross-linking reaction. This shrinkage can be helpful in easing the problems of extraction of the moulded part from the cavity. It is, however, essential to ensure that the mould cavity has suitable draw tapers and good surface finish to avoid sticking problems. When the mould is opened, chilling and thermal contraction of the moulding also assists extraction. For moulding materials such as general purpose phenolics or amino plastics with a moulding shrinkage from 0.5 to 0.8%, adequate ejection from the cavity is possible with a taper angle of 2° on inside surfaces and a minimum of 1° on outside surfaces of the component in line of extraction. For mineral or fibre reinforced phenolics or similar compounds with lower moulding shrinkage in the range 0–0.4%, draw tapers should be increased to 4.5%. These factors, and many others, indicate the importance of both the component designer and the mould designer fully understanding the process and eliminating as many problems as possible at the design stage to avoid more significant and costly problems during subsequent development phases.

5.1.2 Compression moulding presses

Developments in the design, construction, control and ancillary features of compression moulding presses during recent years have done much to enhance the competitiveness of the process and overcome many of the long held weaknesses of this technology.[39]

Many vertical compression presses are of a closed frame construction, as shown in Fig. 5.3, often with means to control deviations from parallelism of the platens under off-centre loading conditions on larger models used to produce close tolerance parts. The alternative column construction presses are sometimes preferred when greater access is required from the sides of the platens. Nowadays it is increasingly common to find presses integrated with robotic handling and other advanced systems (Fig. 5.4).

First of all, a modern press has a fully integrated control system covering not only the actuation steps in the cycle but also direct setting of all pressures, paths, speeds, times and temperatures from a central control cabinet. In this way, and if automatic powder (or pellet) loading and component extraction is employed, the modern compression moulding unit can be considered to be a fully automatic process in the same way as an injection moulding process.

Press specifications tend to fall into two categories. The 'standard' model is likely to have potentiometric parameter setting, PLC or contactor electrical

Figure 5.3 Large all-hydraulic compression press with closed frame construction and microprocessor control system. (Courtesy of Maschinenfabrik J Dieffenbacher GmbH & Co.)

control systems and proportional controlled valve hydraulics. The 'high precision' press is likely to have a microprocessor control system (sometimes in conjunction with a PLC) which allows screen and keyboard setting and monitoring of the process. Velocity and pressure profile setting is possible with the latter, employing appropriate servo hydraulic solutions. As would be expected, input program downloading and process data acquisition, monitoring and control can be included in the specification.

Long stroke presses are commonly specified for general purpose situations, whereas short stroke presses are favoured for large area, shallow mouldings, particularly when using rapid-reacting formulations. A typical

Figure 5.4 Large column type compression press for production of SMC automobile components. (Courtesy of Maschinenfabrik J Dieffenbacher GmbH & Co.)

example of this is the moulding of large panels in a polyester SMC compound where a closing speed of 1000 mm per second with a rapid pressure build up may be required over a relatively short distance to avoid compound pre-cure before the mould has fully closed. With such presses it is also now common to incorporate hydraulic systems and linear transducers to ensure parallel movement of the platens, thus improving the quality and consistency of large area mouldings.

Other ancillary features now available include facilities for fast changing of hot moulds plus a wide range of ejection systems (Fig. 5.5).

The main advantage of the closed frame press construction is in the event of off-centre loading. Lateral rigidity is greater than with presses of column

Among the options supplied by DIEFFENBACHER for moulding work are;

- Auxillary devices such as core pullers, pilot-run pressing equipment with suitable control systems and connections for clients' own handling equipment
- A wide variety of ejectors for use on the presses
- Sliding tables of single- or double-sided pattern

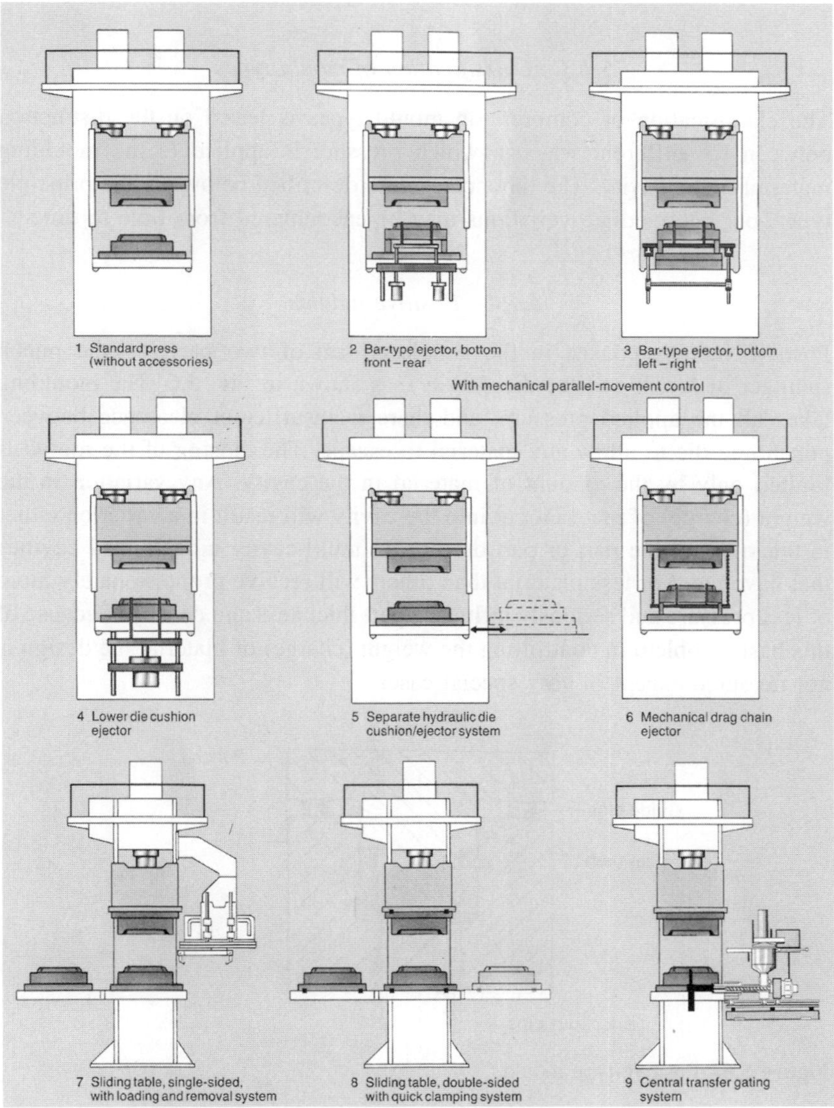

Figure 5.5 Illustrations of a compression press range offering a variety of ancillary features. (Courtesy of Maschinenfabrik J Dieffenbacher GmbH & Co.)

design (approx. 35% higher in x axis and 50% higher in y axis). As mentioned before, the principle advantage of the column construction is better side access which is beneficial when mould changes are complicated, or auxilliary devices are attached to the mould.

5.1.3 Classification of mould types

The classification of compression mould types is based on the distinction between the different ways in which pressure is applied to the moulding material in the cavity. The three categories described below are the principle types, but intermediate variations may be encountered from time to time.

5.1.4 Positive moulds

Positive moulds consist in the simplest form of two parts — the punch (plunger or force) and the die (cavity), as shown in Fig. 5.6. The moulding takes all the applied pressure, and there is insufficient clearance between punch and die to allow any material to escape. The closing of the mould is limited only by the amount of material in the cavity. Any variation in the weight (charge) of material put into the cavity will result in a variation either in thickness of the part or part density. In multi-cavity moulds, any cavities that have more or less material than others will receive proportionately more or less pressure and seriously influence part thickness and density. Because of this basic problem in controlling the weight (charge) of material the design is not favoured except in very special cases.

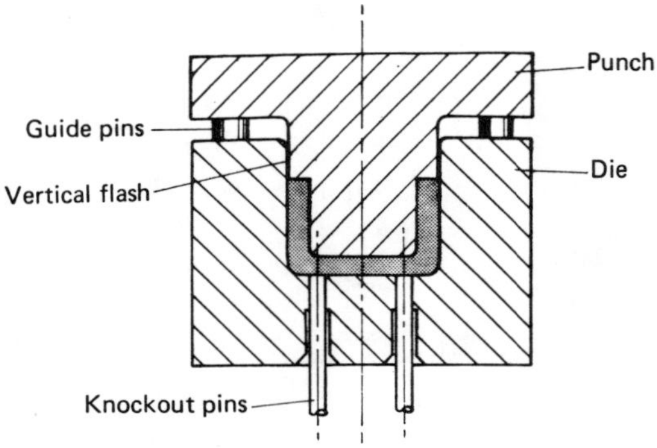

Figure 5.6 Positive-type mould.

5.1.4.1 Semi-positive moulds

These are similar to positive moulds, but the punch does not come down entirely onto the moulding, but to a stop formed by an annular part of the mould surrounding the moulding (see Fig. 5.7). The pressure is therefore

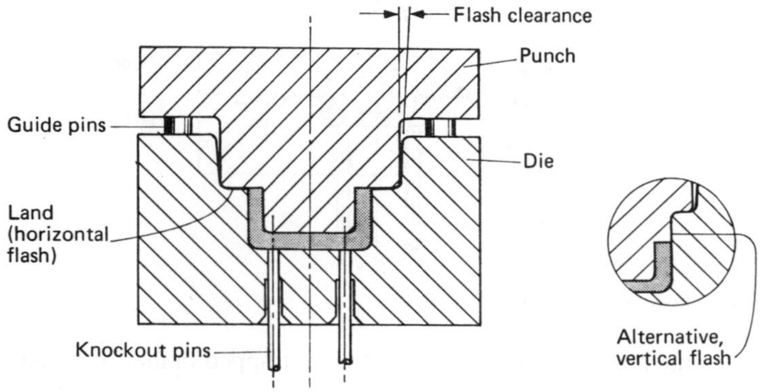

Figure 5.7 Semi-positive type mould.

taken partly by the annular ring, or rather by the flash forced onto it by displacement of the slight excess of powder that has been deposited into the mould cavity.

Mouldings of consistent thickness and weight are quite easily produced. The excess material that escapes as flash is normally only about 2–5% of material used, depending on size, and complexity of flow. Flash thickness can be controlled to less than 0.15 mm. Multi-cavity moulds are a much more viable proposition, and this type of mould is universally used for all types of thermosetting materials. The flash clearance between punch and die is typically 0.15 mm for soft-flow materials and a little more (say up to 0.25 mm) for filled materials. The land areas should not be too wide, 4 to 8 mm being sufficient for most cases.

An alternative to the semi-positive horizontal flash type of mould is the semi-positive vertical flash type. As the description implies, the first type produces flash horizontally and the second type produces vertical flash.

If the part being produced has a height tolerance, then the horizontal flash type is preferable. The surface of the moulding remains untouched when excess flash is trimmed off, thereby maintaining the tolerance. This type is the most popular — with deflashing usually being done by barrelling.

When finishing requirements are less critical, and flash removal from the top face of the moulding is acceptable, then a vertical-type flash mould may be preferable.

5.1.4.2 Flash moulds

Flash moulds are only a simplified form of semi-positive moulds. The mould consists of two parts only, the die and the punch (see Fig. 5.8).

An excess of material is loaded into the die, and the punch brought down onto it. The powder fluxes with heat, and any excess is forced between the

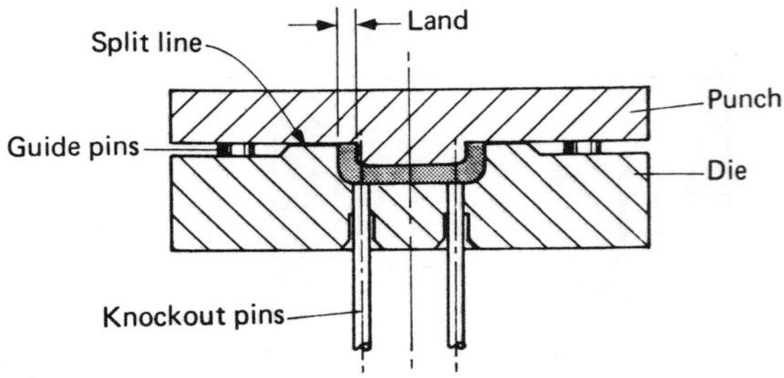

Figure 5.8 Flash-type mould.

faces of the punch and die. Since material is not trapped in the cavity until the mould is almost closed, it is necessary to overcharge the mould, and an excessive amount of material may be wasted.

This type is now used exclusively for tableware products or any large flat moulding of shallow depth, or for multi-cavity button moulds. Close tolerance parts cannot be produced consistently in flash moulds.

5.1.5 Automatic moulding

Automatic compression moulding operations are obviously more attractive than manual and semi-automatic operations when quantities and product life justify the investment and development costs.

It is advisable to specify moulding presses with a fast dry cycle time (typically 3–7 s) and compounds with relatively fast cure characteristics to arrive at overall cycle times that will be economic and justify the cost of the automatic set-up.

Having made this investment, it can be expected that product quality can be higher and more consistent, because of the more controlled and reproducible nature of the process.

If powder is to be used, it is common to employ an automatic powder feed mechanism, as illustrated in Fig. 5.9. Automatic insert loading systems are also becoming increasingiy common, while automated component ejection and unloading mechanisms have been refined over many years. To ensure that automatic ejection can operate in a foolproof manner, it is essential that ejection is engineered to be positive and take place from one side of the mould only. Ejector pins must be sufficient in number and correctly sited to ensure positive lift from the mould, free from stresses that may cause distortion of the product.

5.1.6 Typical compression moulding defects and remedies

Some of the most common defects and suggested remedies are given in Table 5.2.

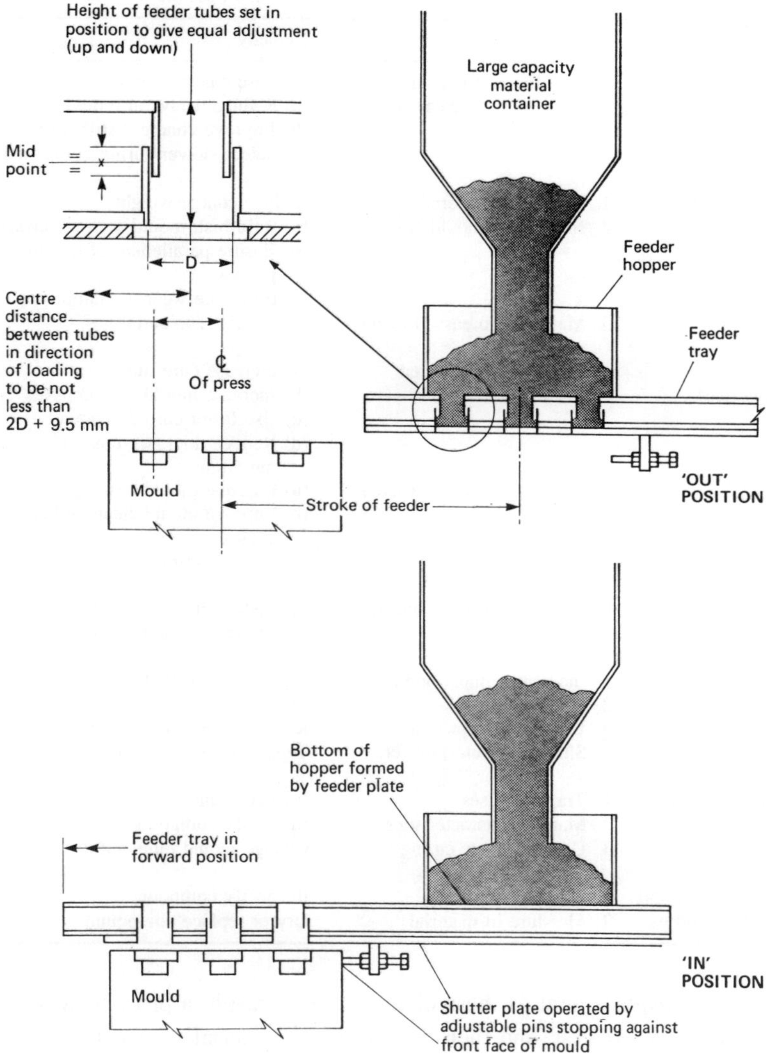

Figure 5.9 Operation of a powder feed mechanism.

5.2 Transfer moulding

The development of thermosetting transfer moulding arose out of the common problem encountered in compression moulding when there is a requirement to produce products incorporating holes or recesses requiring the use of delicate pins or side cores, or when location of inserts in both halves of the mould cavity is required. In these cases the squeezing of material that occurs during mould closure in compression moulding can lead to damage,

Table 5.2 Compression moulding defects — causes and possible remedies

Defect	Cause	Remedy
Short moulding	1. Insufficient material	Increase charge weight
	2. Pre-cure before cavity filled	(a) Reduce delays in mould closing (b) Improve charge distribution (c) Select slower curing compound
Excessive flash	1. Too much material	Reduce charge weight
	2. Inadequate mould closure	(a) Refurbish mould mating surfaces (b) Ensure parallelism of mould and platens (c) Eliminate wear in clamp unit
	3. Material too easy flowing	Select stiffer material
Blisters, porosity and flow marks	1. Moulding undercured	(a) Increase cure time (b) Increase mould temperatures (c) Use faster cure compound (d) Reduce thick sections in moulding
	2. Escape of gases restricted	(a) Improve mould venting (b) Improve mould cleaning between cycles (c) Use breathing sequence
Burn marks	Local degradation of material	(a) Reduce mould temperatures (b) Improve mould venting
Poor ejection	1. Inadequate taper angles	Increase taper angles
	2. Undercuts, etc.	Investigate and eliminate
	3. Unbalanced ejection	Refurbish or modify ejection system
	4. Shrinkage onto pins, etc.	Taper pins or modify ejection system
Staining of mould	1. Trapped gases	Improve venting
	2. Material characteristics	Respecify compound
	3. Unsatisfactory curing	Change mould temperatures
Poor mechanical and electrical properties	1. Incorrect material	Respecify compound
	2. Moisture in material	Dry or replace compound

distortion or displacement of these elements. As a result, a process in which the mould is closed first, and then the moulding material is transferred into the cavity was developed. This process is termed transfer moulding.

5.2.1 Process description

There are two commonly used methods of transfer moulding.

5.2.1.1 Integral pot method

This is achieved by using a mould into which all the facilities for transfer moulding are built, as shown in (Fig. 5.10).

The integral mould is usually a 3-plate mould in which the bottom plate

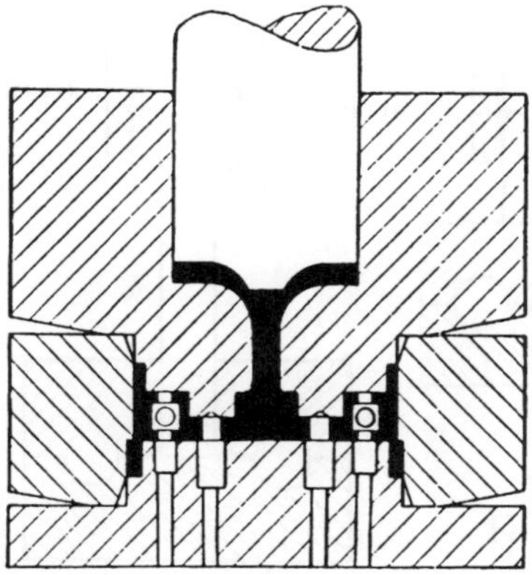

Figure 5.10 Simple integral pot transfer mould.

and the bottom half of the middle plate contain the component forms, runner system, etc. The top portion of the middle plate contains the material chamber (or pot), and the top plate contains the plunger that applies pressure to material in the pot as the mould closes.

The advantage of this type of tooling is the fact that it can be used on standard compression moulding machines without the cost of expensive additional features. The disadvantage is that, for a period of time as the mould is closed, the only pressure keeping the cavity plates closed is that transmitted through the material in the chamber by the plunger. With this situation, back pressure from material in the cavity may be sufficient to cause mould plates to open, with flash or inadequate filling being the outcome. For this reason, the pressure applied by the plunger on the material in the 'pot' must always exceed that in the cavities and this is best achieved by ensuring that the projected area of plunger and pot exceeds the total projected areas of the components and runner system. This excess should be in the order of 10%.

5.2.1.2 Separate pot method

By using a moulding machine where the transfer facility, that is the pot, plunger and appropriate hydraulic pressure circuit, is built into the machine (Fig. 5.11), the mould may be subjected to higher direct clamping pressures.

The pot and plunger may be smaller in surface area provided that it is adequate for the volume of material, and pressure applied to the material may be more related to the total requirements. This method is more desirable for products of greater projected areas, reducing the possibility of flash, and giving better control of dimensions or distortion.

Available transfer moulding machines can be purchased which allow for the transfer of material from the top, bottom or side of a mould, this is

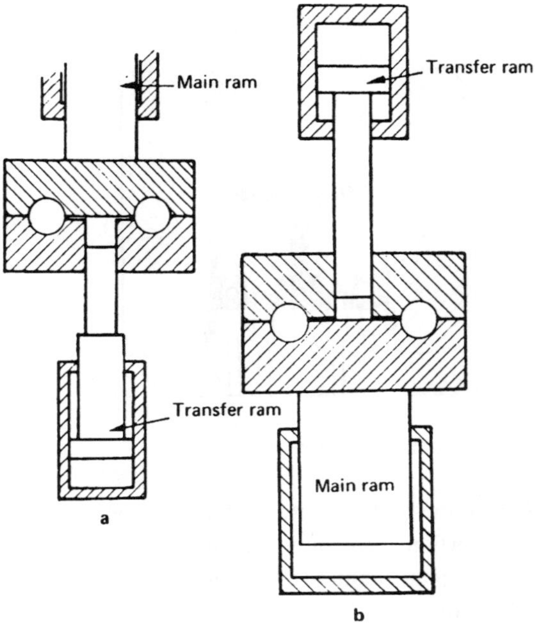

Figure 5.11 Auxillary ram moulds: (a) upstroking, (b) downstroking.

through the mould split line. Further modification allows for the use of screw preplasticizer feeding material into the transfer pot, from where it is transferred by plunger into the cavity. The choice of method of transfer moulding depends upon product shape, method of feed to be adopted, quantity requirements, etc, and will be determined comparative to other methods of production.

Transfer moulding pressures should be calculated on the basis of the total projected moulding area, comprising component(s), runners and pot area. Typical recommended transfer pressures for the main families of thermosettting materials are given in Table 5.3.

Mould temperatures approximate to those used for compression moulding, although it is not unusual to maintain the transfer pot some 30°C lower than the main mould to reduce the risk of pre-cure.

Care must be exercised in the placement of vents at the extremities of flow to facilitate escape of air, moisture and gases arising from the chemical reaction during the passage of material through the cavity.

In general it can be said that venting is much more critical for the moulding processes with significant material movement (i.e. transfer and injection) than in compression moulding.

Face vents on the land area may be typically 0.05–0.075 mm deep and 3–6 mm wide leading directly to atmosphere. In addition venting of gases back

Table 5.3 Typical transfer moulding pressures for main classes of thermosetting compounds

Moulding material class	Pressure (bar)
Phenolic	
General purpose	450–700
Medium impact	550–850
Glass reinforced	700–850
Melamine	
Cellulose filled	550–850
Mineral filled	550–850
Glass reinforced	700–1050
Diallyl phthalate	
Nylon reinforced	550–700
Glass reinforced	700–850
Polyester	
Glass reinforced granular	400–800
Glass reinforced dough	80–600
Silicone	
Mineral filled	350–550
Glass reinforced	350–700
Epoxy	
Mineral filled	400–550
Glass reinforced	300–700

along the ejector pins is encouraged by relieving the mould plate to the rear and/or also putting flats on the sides of the pins to allow flash to escape easily to the rear of the mould.

The ejection system must also be designed to allow easy removal of components, gates, runners and the cull of cured material remaining in the pot, thus allowing minimum interruption to the moulding cycle. Irregular and unpolished mould surfaces can cause flash to adhere, and severely interrupt the cycle for extended cleaning operations.

Gate and runner design is important on two counts. Firstly, the gate and runner play an important part in rapidly heating the moulding material during transfer and thereby promoting a faster cure than would be possible when compression moulding the same part. Secondly, the position of the gate governs the flow pattern of material during mould filling. Because thermosetting compounds include fibrous and particulate fillers, the filling process will result in filler orientation. It can therefore be expected that the resultant mouldings will be anisotropic, with mechanical properties, shrinkages and distortions related to the flow patterns and filler orientations.

This characteristic has been the basis of much research, and a multitude of specialist literature exists for the interested reader. However a very simple illustration of one of the fundamental effects is given in Fig. 5.12 which shows how, with a component that is edge fed, the initial material enters the mould in a continuous fluxed worm which folds and gradually consolidates as packing of the cavity occurs (Fig. 5.12a). Figure 5.12b shows that although the fibre alignment in the compound can be expected to be in the direction of flow at the gate, the folding nature of cavity filling usually ensures that fibre alignment in the bulk of the moulding will often be roughly perpendicular to the direction of filling.

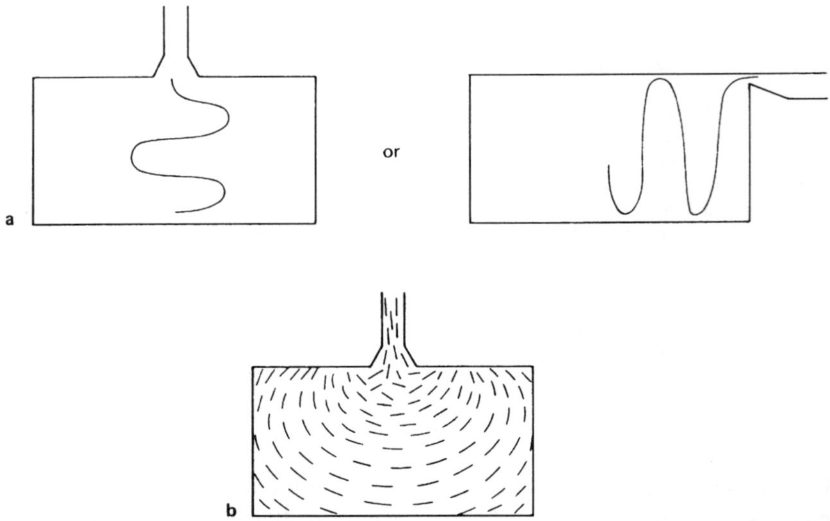

Figure 5.12 (a) Flow fronts during initial mould filling. (b) Fibre alignment during mould filling.

As a general rule the location of the feed point on a typical moulding having length greater than breadth or thickness should be in the centre of the shorter side if warpage is to be avoided. Direct feed of material into the centre of a round moulding (single impression) will usually result in freedom from distortion. Full round runners offer less resistance than half round, square or rectangular sections of the same cross-sectional area. Sharp corners in runners and gates can cause hot spots where material sets prematurely or burns before flowing into the cavities. Multiple gates can produce knit lines where materials flow together. Usually one larger gate is preferable to several small ones.

To minimize required transfer pressures runners should be kept as short as possible. A basic guide for gate dimensions for medium to small components is:

Cross-sectional area of gate (mm^2) = 0.35 × weight of moulding (g)

and

Cross-sectional area of runner = 3 × cross-sectional area of gate

5.2.2 Transfer moulding presses

The design and construction of the transfer press for the normal range of applications is largely based on compression press designs and construction, as discussed in the first part of this chapter.

A typical automatic transfer press for powder or preforms would have a down-stroking clamp unit and an up-stroking integral transfer unit. Figure 5.13 shows the general layout of such a press, showing also an automatic powder feeding and preheating unit and an automatic component stripping mechanism.

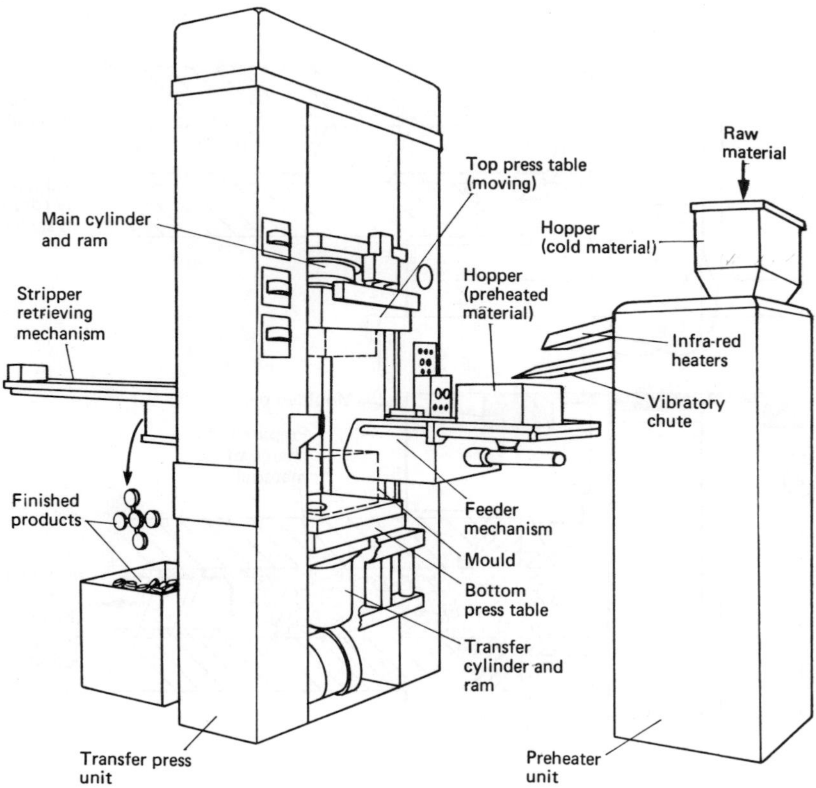

Figure 5.13 Automatic transfer process.

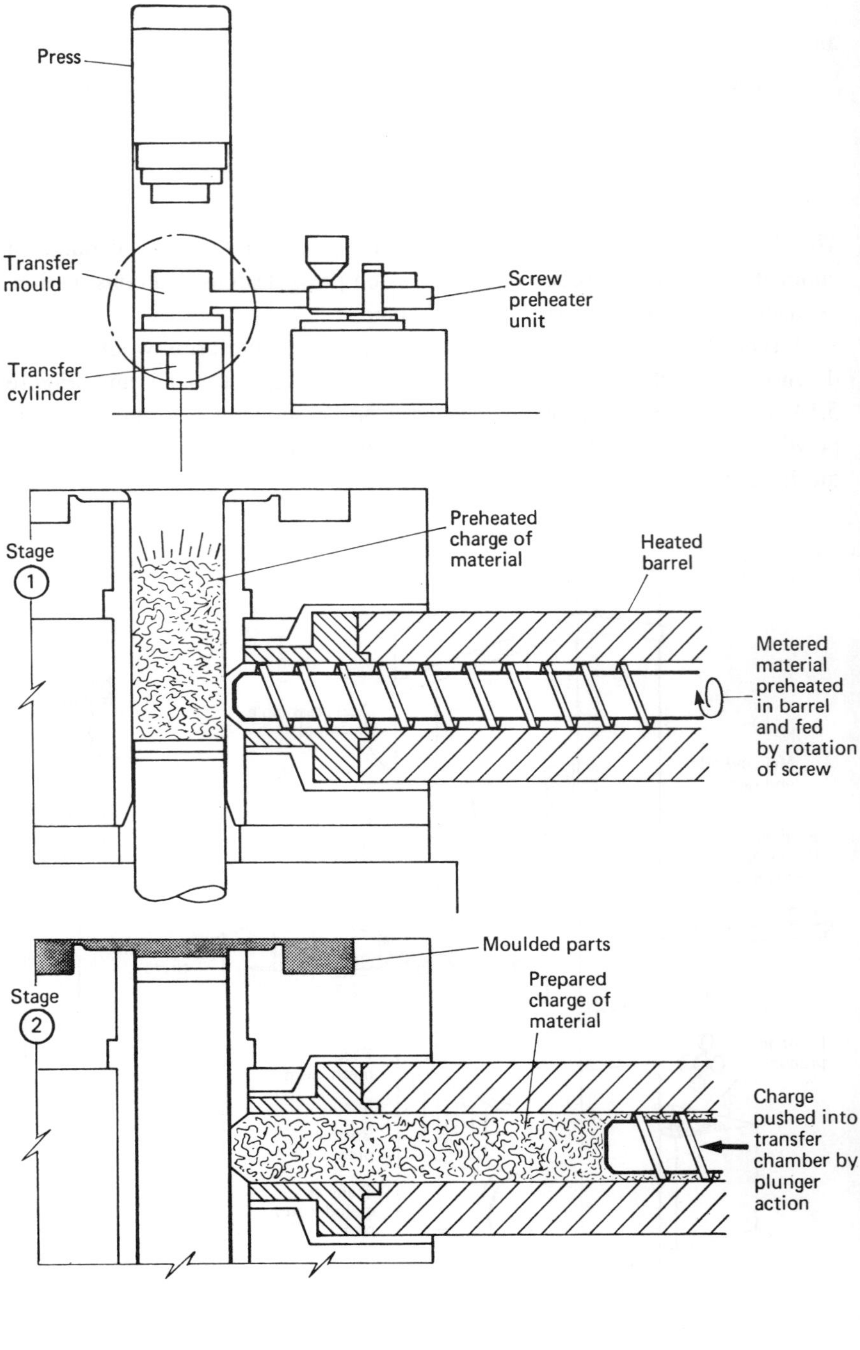

Figure 5.14 Direct feeding pre-heater.

If preheated material is to be used, screw preplasticizing of the compound can be employed as an alternative to conventionally preheated powder or pellets. An arrangement to allow direct feeding of screw preplasticized material into the transfer pot is shown in Fig. 5.14. In some cases a retractable preheat unit is used, which enters the daylight of the press during the open cycle and deposits the plasticized charge into the open cavity of the transfer pot. This latter approach has the benefit of eliminating the need to modify the moulding machine and mould for this type of moulding.

Transfer moulding is still the best method for encapsulating delicate electronic components in low viscosity epoxy and silicone compounds.[40] Special transfer presses are available for this class of work, having refined control of transfer pressures and speeds, and relatively large area platens in relation to clamping force to allow transfer encapsulation of large numbers of semiconductors in a single moulding cycle.

5.2.3 Transfer moulds

The principal parts of a typical mould are shown in Fig. 5.15. In this case standard fixed heating plates attached to the machine platens are used to heat the mould for large moulds, and for long running jobs it may be more common to incorporate the means of heating into the mould.

The design of the transfer chamber is an important feature. lt is common to incorporate a piston groove in the head of the injection piston (or pommel). During moulding, material enters the groove and cures into a piston ring which subsequentiy restricts backflow of material during the transfer sequence (Fig. 5.16).

Care is also necessary in the provision of adequate means of ejection for the components, runner and cull. The need to ensure that positive ejection can take place from one side of the mould has already been discussed. Some times hold pins and/or undercuts are designed into the mould to retain all of the moulded material in the one half on mould opening.

For those readers wishing to go into more detail on these mould design issues, other reference material exists.[41]

5.2.4 Advantages and disadvantages

In comparison to compression moulding, transfer moulding has the following attributes.

5.2.4.1 Advantages

1. Shorter cycle times resulting from friction effects and better heat transfer.
2. Better control of product thickness resulting from a closed mould process.

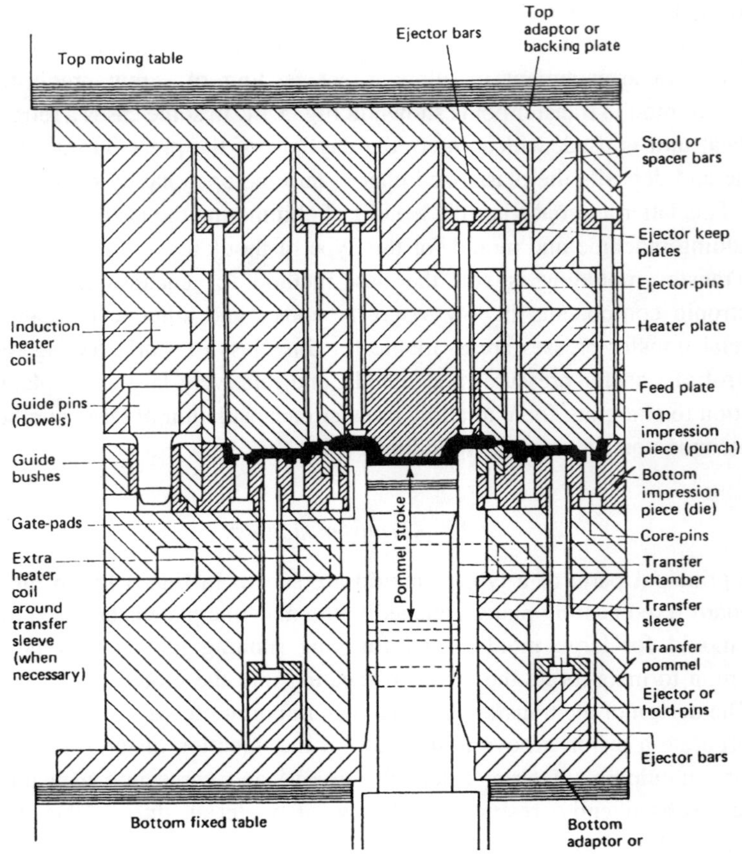

Figure 5.15 Schematic layout of a typical transfer mould.

3. Delicate inserts and mould sections are more easily catered for.
4. Higher tensile and flexural strengths are claimed, resulting from orientation effects, whereas impact strengths may be reduced because more damage may occur to reinforcing fillers.

5.2.4.2 Disadvantages

1. Properties are directional and more susceptible to weld line effects.
2. Variable shrinkage resulting from material flow. Increased tendency for warpage and distortion.
3. Electrical insulation properties of the product may be more difficult to achieve and maintain.
4. Potential for more material wastage in cull and runners.

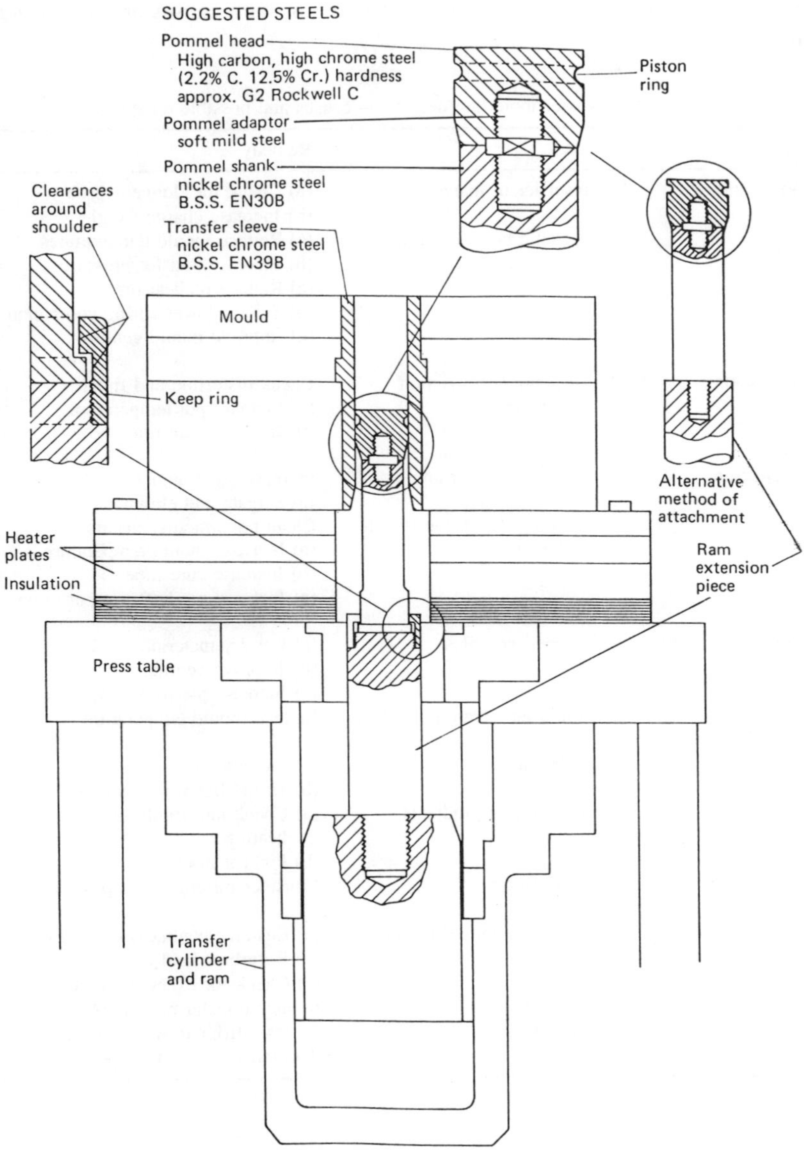

Figure 5.16 Design of transfer chamber and piston (pommel).

5.2.5 *Typical transfer moulding defects and remedies*

The most common defects are listed in Table 5.4, with causes and suggested remedies.

Table 5.4 Transfer moulding defects — causes and possible remedies

Defect	Cause	Remedy
Short moulding	1. Insufficient material	(a) Check for clogged sprue (b) Increase charge weight
	2. Pre-cure before filling	(a) Reduce mould temperatures (b) Increase transfer pressure (c) Reduce preheat time (d) Select slower curing compound (e) Improve mould venting
Sprue sticking	1. Mechanical restriction	Check distortion and finish
	2. Undercure	(a) Increase pot temperature (b) Increase cure time
Poor ejection	1. Inadequate taper angles	Increase taper angles
	2. Undercuts, etc.	Investigate and eliminate
	3. Sticking of pin, inserts, etc.	Clean pins, modify inserts
	4. Undercure	(a) Increase mould temperature (b) Increase cure time (c) Increase pre-heat time or temp
Blisters and porosity	1. Trapped volatiles	(a) Pre-dry material (b) Improve venting (c) Increase pre-form temperature
	2. Mould temperature too high	Reduce mould temperature
Warped products	1. Undercure	(a) Increase cure time (b) Use stiffer flow material
	2. Temperature gradients	(a) Check uniformity of mould heating (b) Preheat inserts
	3. Flow effects	Consider different gate position
Excessive flash	1. Inadequate mould closure	(a) Ensure cleanliness of land faces (b) Refurbish land faces (c) Check for excessive wear
	2. Undercure	Reduce transfer pressure & speed
	3. Material flow	(a) Use stiffer flow compound (b) Increase mould temperature

5.3 Preforming and preheating

Preforming (or pelleting) is a convenient way of ensuring a consistent charge weight of granular moulding compound and is used for compression or transfer moulding. It also greatly facilitates the preheating process.

Preforming machines are available in various sizes and described by their nominal clamping tonnage, e.g. 35 tonnes, 70 tonnes.

Cylindrical tablet preforms should have a height of not more than 50% of the tablet diameter. In some cases it will be necessary to employ a multiple of preforms to load a cavity rather than attempt to use a single large preform. The tolerances to which a preform weight can be controlled varies with the type of compound and its granulation. However a 2.5 per cent variation is common. It should be borne in mind that the preforming operation can sometimes cause some attrition of fibre reinforcement within the moulding compound, resulting in reduced mechanical properties in the product.

In general, thermosetting moulding materials are formulated to ensure that very little curing takes place below, say, 60°C to ensure reasonable shelf life, The purpose of preheating is to reduce moulding cycle times by reducing the material heating element of the moulding cycle. Since preheating to temperatures higher than 60°C will initiate the curing reaction, it follows that the delay between the preheating operation and introduction into the mould must be small and controlled.

The most common method used in commercial moulding is high frequency dielectric preheating. A dielectric preheater (Fig. 5.17) consists of two plate electrodes connected to an electric current such that they are of opposite polarity. The circuit normally consists of a power transformer that converts the mains voltage to a high voltage (say 6000 V), a rectifier to convert AC current to pulsating DC, and a thermionic valve. The system is such that the polarity of the electrodes is reversed and thus develops an alternating current, whose magnitude depends on the frequency and potential difference applied. Typical operating frequencies are 50–100 MHz.

The material to be preheated acts as a dielectric between a parallel plate condenser. The alternating current causes molecular agitation and heating within the compound. Since the temperature rise is rapid and consistent throughout the compound, precure is kept to a minimum. The rate of heating is dependent on the following factors:

Voltage
Frequence
Density of material
Specific heat capacity
Power factor
Relative permittivity

The theoretical rate of heating can be calculated from the classical physics formula:

$$\text{Initial temperature rise (°C)} = \frac{\text{heat generated per second}}{\text{mass} \times \text{specific heat capacity}}$$

where heat generated per second is equivalent to power dissipated. The

Figure 5.17 6 kW, 80 MHz preheater. (Courtesy of Radyne Ltd.)

method for calculation of the power dissipated is relatively complex and can be examined elsewhere.[41]

Preheating of moulding powders, rather than preforms is often carried out using infra-red lamps in conjunction with a vibratory feed arrangement. Capital expenditure is lower than for preforms and dielectric preheating, but the efficiency and control is not as effective.

Hot air oven (convection) preheating is sometimes used, particularly with urea–formaldehyde compounds. However, this method has a greater tendency to cause stiffening of the moulding compound, due to the loss of moisture during this process.

Finally, as discussed in the section on transfer moulding, screw preplasticization can be a very effective alternative, since it probably produces the best and most uniformly preheated charge, and can be automatically linked into the moulding sequence.

CHAPTER 6

Injection moulding of thermosetting plastics

6.1 Introduction

For many years it was assumed that the irreversible curing reaction undergone by the thermosetting plastics rendered them unsuitable for injection moulding by direct screw transfer. It was only during the early 1960s that experimental work in Japan and western Europe showed that screw injection techniques could be usefully adapted to handle this family of materials. These developments at once provided the opportunity for thermoset moulders to streamline process routes, by linking the benefits of automatic material feeding and processing to the possibility of substantial cure time reductions often achieved by the process.

The net result has been to provide a massive stimulus to the thermosetting side of the industry, which otherwise might have declined to only specialist and low volume applications. In many instances, the use of thermosetting injection moulding has resulted in applications previously lost to the more rapidly processed thermoplastics, once more re-establishing themselves in thermosetting moulding compounds at a competitive piece price.

Today, thermosetting injection moulding is becoming increasingly accepted as a routine moulding operation. It can be found in general moulding shops alongside thermoplastics operations and not just in use within specialist enterprises.[42,43]

This coming of age is very heartening, but it makes it all the more important that the engineers involved, from design to manufacture, appreciate the different nature, characteristics and requirements of this technology. While some rules for thermoplastics injection moulding still apply, many are fundamentally different.

6.2 Characteristic features of the thermoset injection moulding process

6.2.1 Process description — granular compounds

Granular thermosetting materials, such as the phenolics, urea–formaldehyde and melamine–formaldehyde compounds are suitable for gravity feed hoppers mounted on the injection cylinder in the conventional manner (Fig. 6.1).

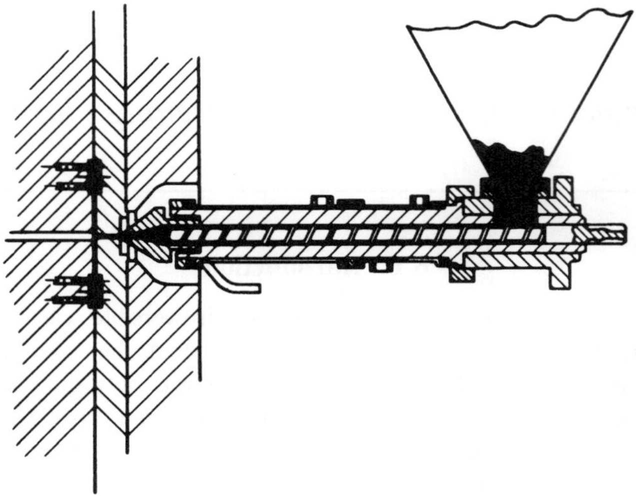

Figure 6.1 Injection cylinder and gravity feed hopper details of a thermoset injection moulding machine.

As with thermoplastics, we require to feed the moulding material along the screw (with the screw moving backwards in the barrel) and at the same time transform the granules into a plastic melt which accumulates at the front of the cylinder in the nozzle area. Conducted heat from the barrel and frictional energy from the screw achieve this fluxing. In the case of thermosetting compounds strictly limited amounts of conducted heat are required to assist the fluxing of the moulding material in the barrel, a major contribution coming from frictional energy from the screw plus the initial stages of the exothermic reaction within the compound. Consequently, it can be seen from Fig. 6.1 that the barrel of the thermoset injection moulding machine has a recirculating hot water (or oil) heating jacket system.

A fluid heating jacket is preferred to a system of electrical resistance (band) heaters, since there is little need for a steep temperature gradient to be applied externally along the barrel. Rather, it is necessary to ensure overall consistency of temperature and to maintain temperature stability against external disturbances or internal temperature fluctuations that could provoke precure of material in the barrel.

As with thermoplastics moulding, the amount of frictional (shear) heating is controlled by the application of an adjustable back pressure on the screw, which opposes the backward movement of the screw during the shot preparation phase of the cycle. The volume of melt created at the screw tip (i.e. shot size) is also controllable by measuring and limiting the screw backwards movement by either an adjustable limit switch or a displacement transducer. During this feeding stage it is normal for the transition from granular material to a soft putty-like melt to occur approximately 2–3 flights back from the nozzle end of the screw. An important side effect of the compression and shearing of the melt in the barrel is the escape of gases, moisture and other volatiles in the reverse direction along the screw. Consequently the screw preplasticizer provides an unrivalled method of preparing an evenly plasticized, homogeneous melt, free from trapped gases and moisture.

It is normal to operate without an external heating source on the nozzle. Usually the controlled exothermic reaction in the moulding compound is sufficient to maintain the satisfactory working temperature of the nozzle. Contact with the hot metal of the mould also provides a contribution. In cases where very fast-curing compounds are being processed, or with larger machines, it is sometimes necessary to provide a separate temperature-controlled fluid jacket for the nozzle to ensure stable processing conditions.

As with compression and transfer moulding, it is necessary to use a heated mould, suitably insulated from the machine platens, to complete the curing of the moulding. Electrical cartridge-type heaters are widely used for mould heating, largely for reasons of flexibility and ease of maintenance.

6.2.2 Process description — bulk (dough) and putty compounds

Styrene-modified polyester bulk moulding compounds (BMCs) and polyester alkyd putty compounds present special handling difficulties for injection moulding. Since these compounds are based on liquid resin systems, a force-feed hopper system can be employed to feed material onto the screw through the feed throat in the barrel.[44] Alternatively, in the case of putty compounds and also polyester sheet moulding compounds (SMCs), a preshaped tape of moulding material can be made to feed from a drum into the throat of the barrel.

An illustration of a typical force-feed hopper is shown in Fig. 7.11. It has been found that greatest feeding and shot weight consistency is achieved when the residual material cushion in the feeding hopper is not allowed to reduce below 30% of the hopper volume, the material feed temperature is controlled, and a closed loop control of stuffing pressure is instituted.[45]

An alternative approach to the problem of injection moulding of BMCs is to dispense with the screw preplasticizer system of the conventional injection moulding machine, and instead use a system of rams to fill the injection cyclinder and then inject the mould. Such a system is shown schematically in Fig. 6.2. Such plunger systems normally require a shut-off valve to be fitted to the nozzle to avoid extrusion of the dough during filling of the injection chamber.

As might be expected, injection units are also available which combine the elements of screw and plunger. In this case a screw transfer and hopper unit is mounted above a horizontal plunger unit (Fig. 6.3) such that the screw meters

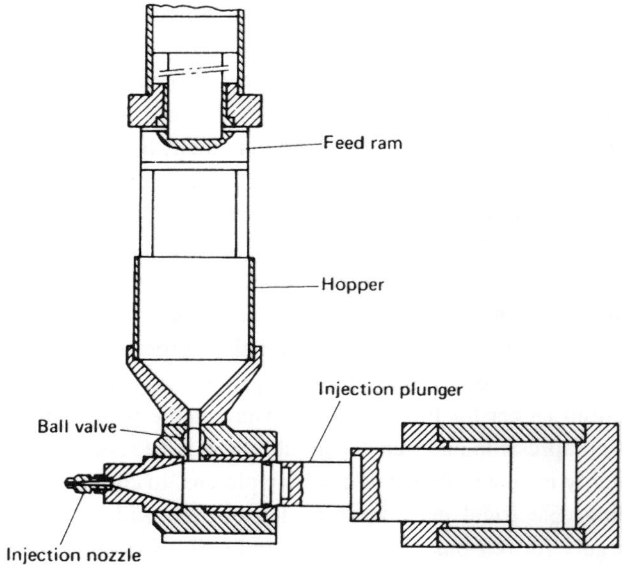

Figure 6.2 Double plunger injection unit for bulk moulding compounds (and rubber).

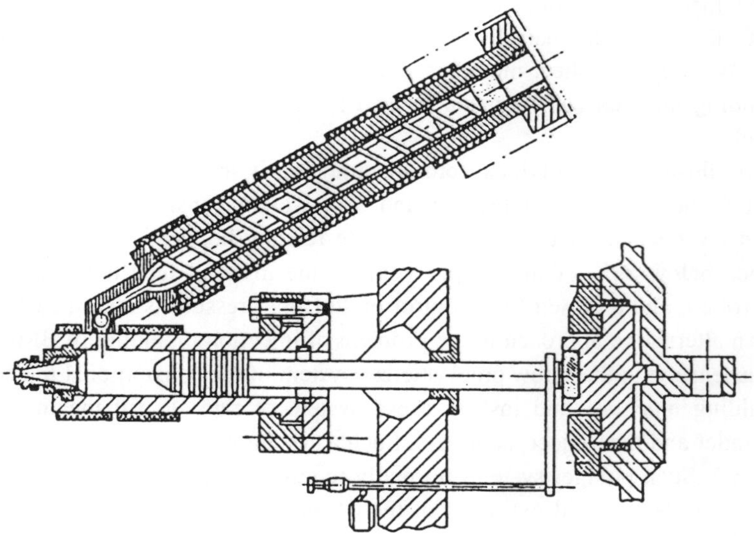

Figure 6.3 Example of a screw feed–plunger injection unit. Reproduced by kind permission of Carl Hanser Verlag from **Johannaber**, *Injection Moulding Machines (3e)*, Carl Hanser Verlag, München-Vienna 1993.

and feeds the required charge down into the throat of the main injection barrel. It is claimed that this system offers the best of both worlds preserving the accuracy and ease of a screw system to obtain high metering accuracy, while a plunger system for direct injection is claimed to minimize fibre degradation in the final part and reduce the number of expensive components subject to wear in the injection unit.

The moulding and performance of polyester BMCs and SMCs is covered in greater detail in Chapter 7 of this book.

6.3 Thermosetting injection moulding machines

The majority of thermoset injection moulding is carried out using screw injection moulding machines, which have many features in common with thermoplastics equipment. It is therefore quite usual to find that a machinery manufacturer will offer a basic machine which can be specified in either thermoplastic or thermosetting format.

It is not the purpose of this book to provide an in-depth treatise on the design and construction of the in-line screw injection moulding machine, which is adequately covered elsewhere.[46,47] Comment is, however, made on the most pertinent aspects of the moulding machine that relate to the processing of thermosetting compounds.

The special requirements for feeding and preparation of the thermosetting melt in readiness for injection into the hot mould have already been described in the previous section of this chapter. In the case of the conventional in-line screw injection unit, the screw is likely to have less flight depth and a smaller channel–depth ratio than typical screws for processing thermoplastics, typically a channel–depth ratio of 1:1 to 1:1.3. This approach ensures that undue shear heating of the material, with risk of prepolymerization in the barrel, is avoided. Non-return valves are normally not required. Although this may result in some backflow of material over the screw flights, it is still normal to use a shorter screw than for thermoplastics, with 12–14 length/diameter ratio being typical. Screws with a small increase in flight depth at the front generally have less back flow and therefore a reduced tendency to screw flight wear. Screws are made of fully hardened steel of high toughness. They may sometimes be chromium plated. It is common for the barrel to have a corrosion and abrasion resistant liner (e.g. an alloy based on Ni–Cr–Co–W–B). Nozzles, of the simple open variety, may be made from stainless or stainless ionitrided steels.

As the reader may know, the (mould) clamping unit of an injection machine may be either a mechanical (toggle operated) positive locking system, a hydraulic force locking system or a mechanical–hydraulic force locking system. It has become widely accepted that hydraulic clamping systems are preferable for thermosetting work, mainly because the expansion

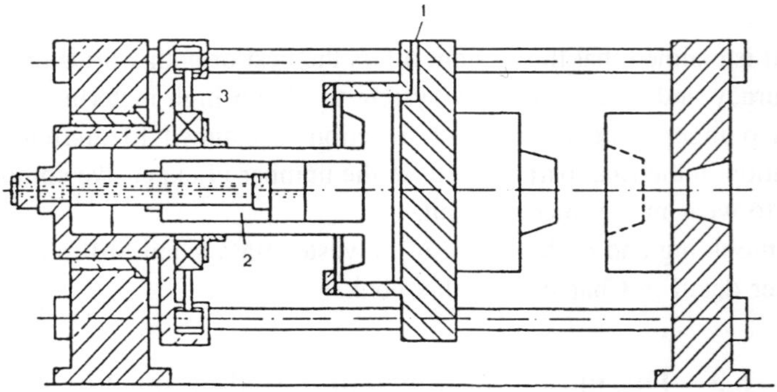

Figure 6.4 Hydraulic clamping unit with auxillary fast closing cylinder. (1) Main clamping cylinder, (2) fast closing cylinder, (3) locking cylinder. Reproduced by kind permission of Carl Hanser Verlag from **Johannaber**, *Injection Moulding Machines (3e)*, Carl Hanser Verlag, München-Vienna 1993.

of heated moulds is automatically catered for when a hydraulic system is used, and clamping forces can therefore be guaranteed at all times. Although speed of motion of a toggle system is still greater, hydraulic clamping systems are now more comparable, and they do allow much more flexibility in programming clamping speeds and positions. The hydraulic unit will often have a high speed auxiliary cylinder to provide rapid closure, with a large central hydraulic cylinder to provide the main clamping force (Fig. 6.4).

To achieve the necessary accuracy and repeatability in machine function it is essential to have a precise and reliable hydraulic system through which the required process pressures and speeds can be controlled. There are three common basic versions likely to be encountered:

- Digital hydraulic units
- Analog-based hydraulic units with proportional valves
- Fully closed-loop servo-controlled systems

A typical example of a digital hydraulic block is given in Fig. 6.5, where the required flow rate can consist of a combination of smaller component flow rates, the magnitude of each being determined by the diameter of a fixed orifice. Thus many small steps in required flow can be achieved from a relatively few fixed output valves.

Analog hydraulic systems incorporating solenoid-operated proportional valves, capable of high standards of performance, provide a very cost efficient solution. In most cases, these valves are calibrated to allow direct setting from the machine control system.

The third approach of full servo control systems relies on feedback sensors on the machine to measure hydraulic performance in terms of achieved pressures and velocities, and these values are used to apply immediate and automatic corrective action through the servo valves should the actual value

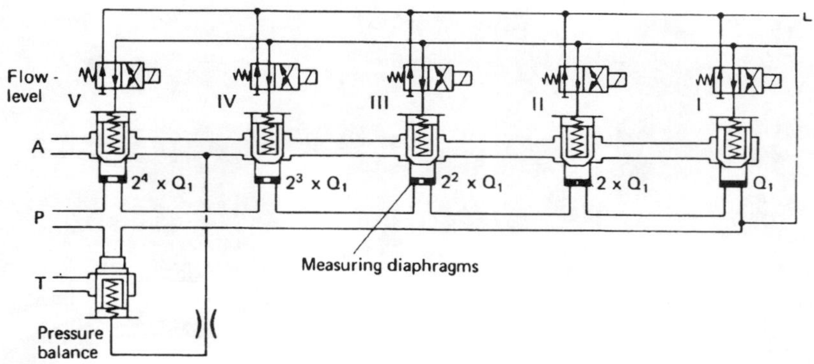

Figure 6.5 A digital hydraulic block for control of flow rates with compensation by pressure regulator. (Courtesy of Demag Kunststofftechnik.)

differ from set value. Servo systems obviously offer potential for greatest precision and control, but their additional expense, complexity and need for higher maintenance standards can sometimes weigh against their use.

It is common practice to employ variable delivery pumps to reduce energy consumption and therefore reduce running costs.

Today, the vast majority of injection moulding machines have a relatively sophisticated microprocessor or multi-processor control system incorporating comprehensive information display from a CRT, liquid crystal or other suitable screen technology. This is usually linked to an input keyboard. Machine setting and interrogation for a wide range of queries can be accommodated in 'stand alone' mode through the keyboard. In addition, it is now common for the manufacturer to offer a wide range of hardware and software to allow downloading and feedback of programs and information to support a range of requirements including central machine setting, collection of production data, process monitoring, quality monitoring, statistical process control, diagnostic/fault finding, etc.

The increasing power and usage of computer technology has, as in many other fields, been one of the dominating changes in the development of the injection moulding process in recent years. There is every indication that this trend is set to continue, with the increasing emphasis on product quality monitoring systems. A typical modern thermosetting injection moulding machine with freestanding control console is illustrated in Fig. 6.6.

6.4 Process setting and control

If one allows for the complexity of the thermoset injection moulding process, the nature of the chemical reaction, and the considerable number of variables to be set and controlled, it is clear that some strategy is required to manage the process and achieve consistent results.[48]

Figure 6.6 Modern thermoset moulding machine with hydraulic clamping and microprocessor control. (Courtesy of Fahr Bucher GmbH.)

It is important to develop a clear understanding of the levels of control possible, what each level entails, in terms of investments and operating requirements, and what benefits and restrictions relate to each category. An example for approaching such a strategy is given in Fig. 6.7, which shows the principle involved in tackling a basically uncontrolled moulding process and systematically moving along a path leading to higher levels of control, and if necessary, moulding shop integration.

What is necessary, and often difficult to establish, is a strong and reliable connection between a relevant and easily measured feature of the product such as weight, a key dimension, or surface finish, etc. and the most influential process control parameter (pressure, temperature, injection speed, etc.).

It is often desirable to employ statistical process control (SPC) techniques to chart the average and range of variance in the measured product feature.[49] This information may be of little use if the link with process variables necessary to allow manual or automatic correction of the process has not been established. In many cases the installed hydraulic and electronic systems in the modern injection moulding machine provide a perfectly adequate level of self-regulation of the process, thereby severely restricting gross variations in the product. However, in cases where extreme control of dimensions, surface finish or similar properties is required, further investigation may be required to establish the most sensitive control measures. In some cases it may be

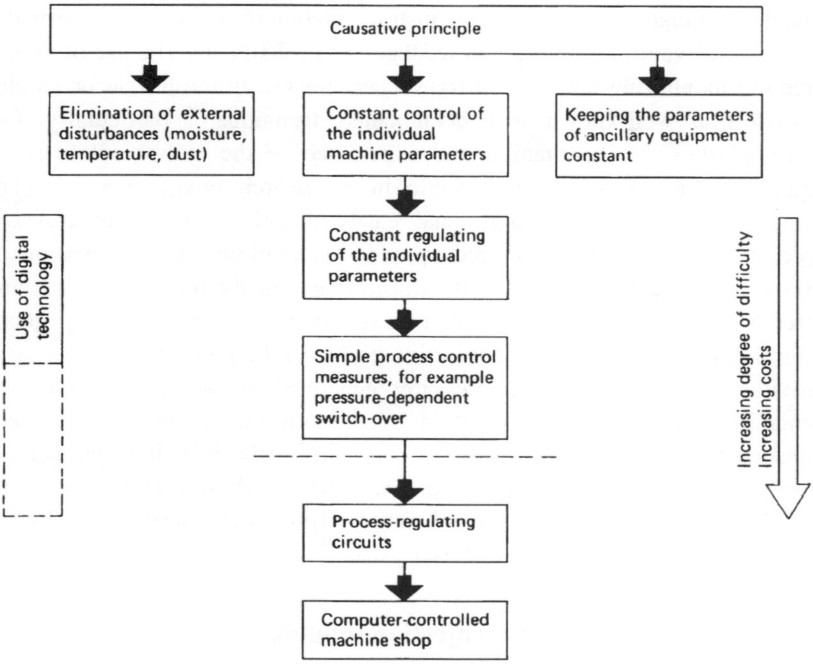

Figure 6.7 A basic control strategy. (Courtesy of Demag Kunststofftechnik).

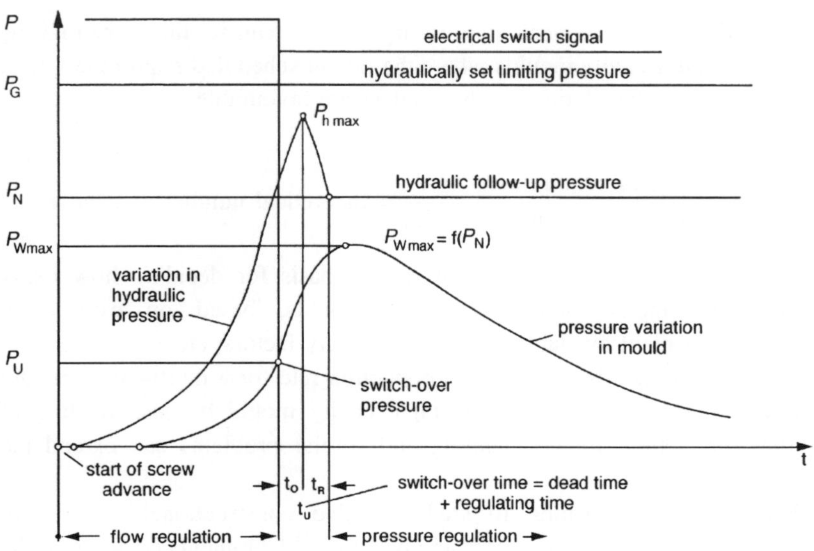

Figure 6.8 Illustration of the different behaviour of hydraulic and melt (cavity) pressures during injection, when a pressure-dependent switchover from flow regulation to pressure regulation is exercised.

found beneficial to resort to further instrumentation of the process to provide the most relevant monitoring and feedback control signals. The use of melt–pressure monitoring and control transducers for use in the nozzle or mould cavity is a well-known and documented technique,[50] particularly for thermoplastics. In the opinion and experience of the author, while such equipment can prove most valuable in a research environment, it can sometimes prove to be fragile and easily disturbed or damaged when specified for routine use in a typical production moulding shop environment. Pressure transducers installed in the injection end of the machine hydraulics offer a more robust long-term alternative. These will provide very good information on the feeding and injection phases of the process, but obviously provide little information on pressures generated in the mould cavity. A generalized graphical representation of the information obtainable from transducers recording either hydraulic pressure in the injection cylinder or cavity (melt) pressure is given in Fig. 6.8, which clearly illustrates this point. For those who wish to understand and interpret these pressure traces in greater detail other reference material exists.[46]

6.5 Injection moulds

One of the key benefits of the injection moulding process is the potential for high manufacturing output by virtue of its speed, repetitive nature and automatic sequence.

It is therefore a logical step to try and maximize this potential by specifying multi-cavity moulds where the rate of scheduled requirement (e.g. parts/week) and the lifetime of the product are favourable.

A simple, rule of thumb, calculation is:

$$\frac{\text{rate of requirement}}{(\text{cycles/hour}) \times (\text{hours available})} = \text{theoretical number of cavities}$$

This number of cavities is then used as a basis for deciding how many cavities per mould and how many moulds, based on experience, practicability, plant available, economics, safety factors, etc.

It must be remembered that the production rate for a multi-cavity mould is rarely a simple multiple of a single cavity mould because feeding of cavities, extraction, mould cleaning and similar problems can extend the cycle.

For the reader wishing to study detailed constructional features of injection moulds, specialist works are available.[51] Comments on some of the more important aspects specifically relevant to thermosetting moulds are provided below.

6.5.1 The feed system

A characteristic runner and gating system is shown in Fig. 6.9. Runners in general, should not be thinner than the thickest section of the moulding and should be as short as possible to avoid problems of excessive shrinkage in the moulding, short mouldings or flashing resulting from the use of high injection pressures. It is normal practice to have a smaller cross section for branch runners than for the main central runner. Because the moulding material becomes more viscous as it flows, it is advisable to have generous radii on the entrance to branch runners to ease change of direction of the flowing melt.

It is considered good practice to feed into the thick portion of a moulding if at all possible.

The best shape for gates is round followed by square (see Fig. 6.10), but these ideal shapes are not always possible. It has been established in practice

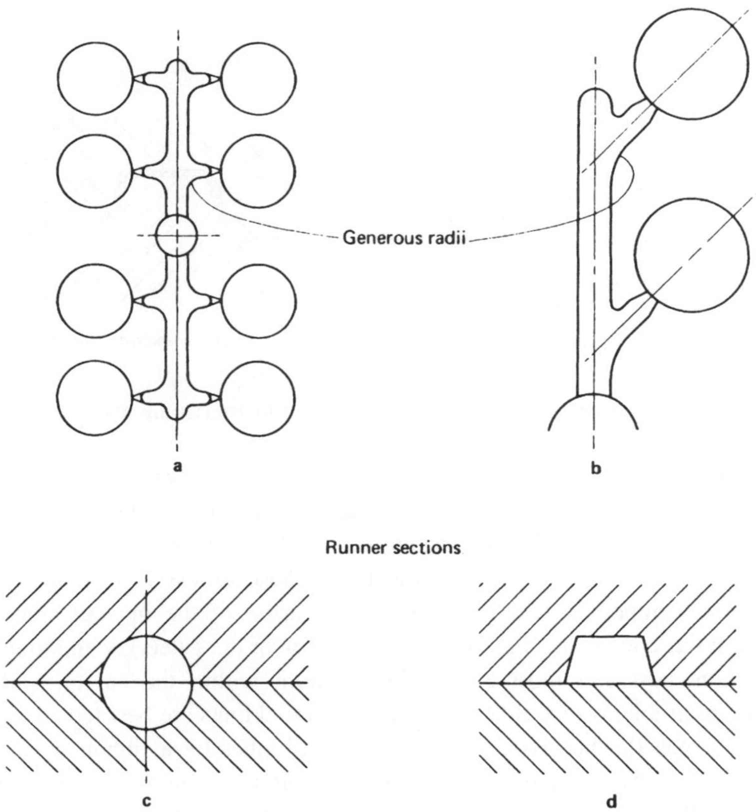

Figure 6.9 Typical examples of runner design. (a) Single runner, suitable for small components; (b) angled branch runner, suitable for larger components; (c) round (conventional) section; (d) trapezoidal section.

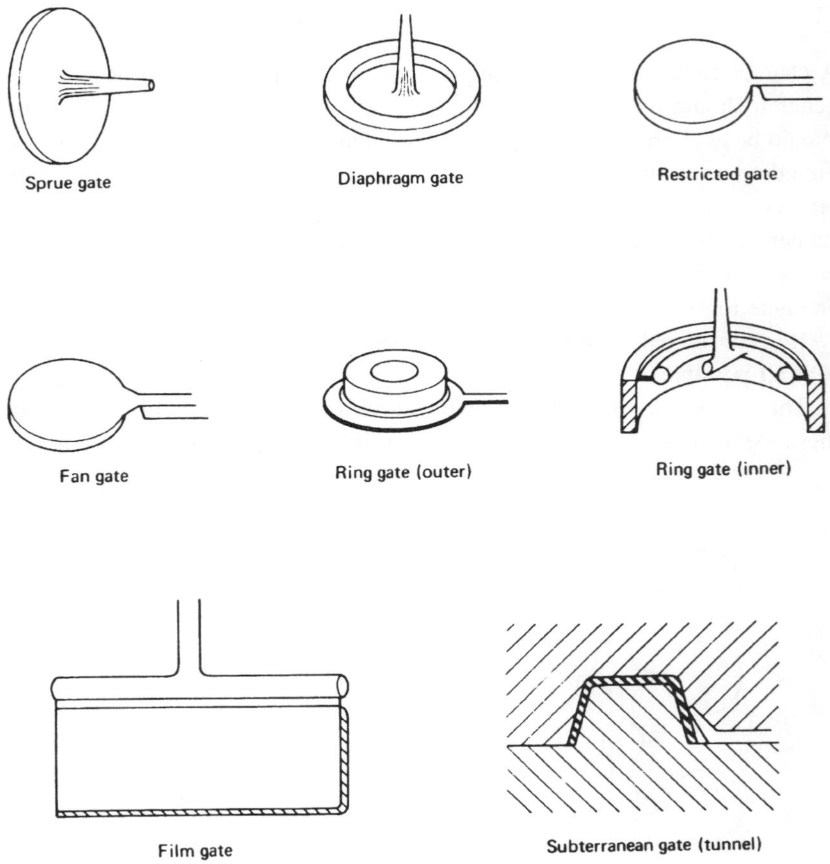

Figure 6.10 Gate design for injection moulding of thermosetting materials.

that with granular thermosetting materials having fine fillers a gate thickness of 0.25–0.50 mm as minimum can be used, but with coarse fillers about 0.75–1.0 mm is usually the minimum size to start with. A gate that is too small can give a great deal of trouble because it can cause excessive heating of the moulding compound, with subsequent pre-curing of the injected material before it has consolidated in the cavity. Position of the gate(s) is important to encourage best and most even filling pattern within the cavity, and to minimize problems of weld lines. Long thin mouldings are best fed from the end, whereas flat box-type parts are best fed either by a ribbon-type feed (film gate) or by direct sprue central gating. Subterranean gates can be used successfully with thermosets, but the angle subtended between the draw line and the tunnel gate should not exceed 30°C to avoid extraction problems.

With a multi-cavity mould it sometimes proves necessary to balance the gates in such a way to ensure that all cavities fill together. In the past, this has always been done by trial and error methods needing much time and experience. The development of computer-based mould-filling simulation software is likely to play an increasingly important role in optimizing runner and gate design for thermosetting as well as thermoplastics moulds. This technology is described in greater detail in section 6.6.

6.5.2 Mould heating and temperature control

As with transfer and compression moulds, heating can be by gas, steam or electricity, with electrical heating by cartridge type heaters being most popular. As a general guide, an allowance of 0.25 to $0.5\,\mathrm{W\,cm^{-3}}$ of all uninsulated steel can be used to calculate total installed heating capacity.

It is necessary to install thermocouples in the mould to allow accurate monitoring and control of mould temperatures.

There are, in the main, two schools of thought and some controversy regarding the best positioning of thermocouples. They can be placed close to the heating source to get a quick response, thus minimizing risks of ‘hunting’ effects. The alternative approach is to site the thermocouples as close as possible to the surface of the mould cavity, regardless of the distance from the heat source.

With the trend towards higher levels of process instrumentation and control, it is now becoming much more common to combine both approaches in the same mould, using the ‘deep’ thermocouples (near the heat source) for control and regulation of heating, while using the ‘surface thermocouples’ principally for monitoring and establishing the need for fine adjustments to mould temperature.

6.5.3 Ejection

Many of the principles for ejector pin design described for transfer moulds in the last chapter are equally valid for injection moulds. In particular, the need for relieve pins to the rear of the mould to avoid seizure resulting from progressive resin build up in the hole is most important (Fig. 6.11).

In any case, such mould ejector systems form an important part of any mould maintenance programme. The benefit of the relieved pin is to offer the potential of extending the useable period between maintenance.

Ejection can be assisted by the use of stripper boards, which can be automatically actuated within the moulding cycle. Mould venting requirements also follow the general recommendations for transfer moulds. Venting down ejector pins, or through sintered porous plugs where moving pins are impossible, can be used to vent to the rear of the mould. Die face venting positions are usually confirmed by trial and error during mould development. Detailed illustrations of typical venting features are shown in Fig. 6.12.

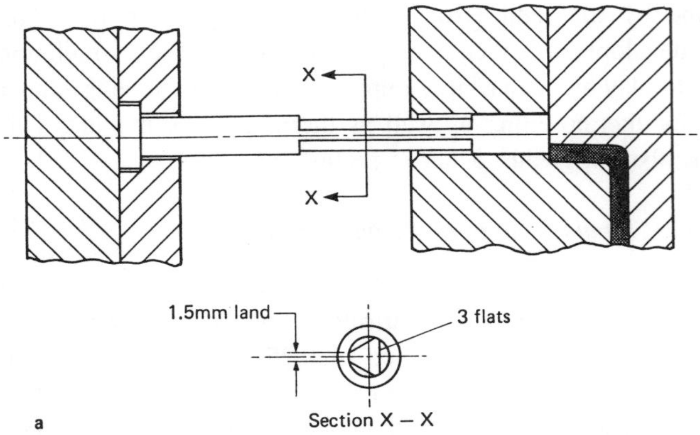

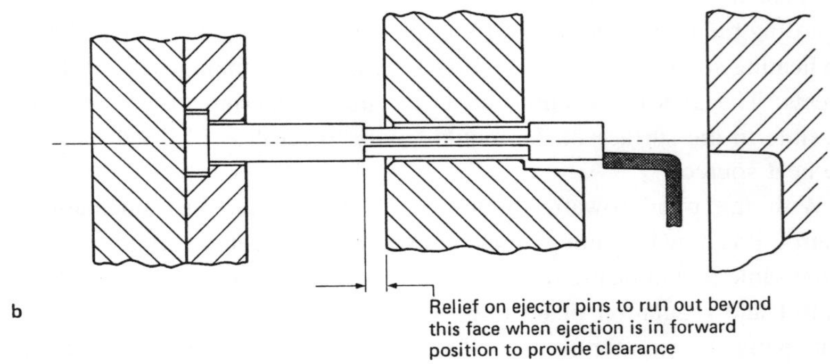

Figure 6.11 Flatted ejector pins to avoid mould seizure. (a) Mould closed, (b) mould open.

In some special cases vacuum extraction is applied to the mould. Theoretically, evacuation of the air from the mould prior to ingress of the melt should be the best aid to mould filling. However, in practice, the use of such systems is not without difficulty. Maintenance of such systems can be difficult in routine production, particularly with respect to the interference of flash contamination in the system. Cycle times can become extended, and there are of course additional equipment costs.

A sectional view of a typical thermosetting injection mould, in this instance for multi-impression moulding of electrical insulator base components, is shown in Fig. 6.13. The main constructional elements of the mould are clearly shown, with the left-hand side of the drawing taking a section through the components, while the right-hand side is taken through the runner. In this application, where the component is expected to withstand heavy electrical loads in service, care would normally be exercised to ensure very adequate venting, along ejector pins and through die face vents, to

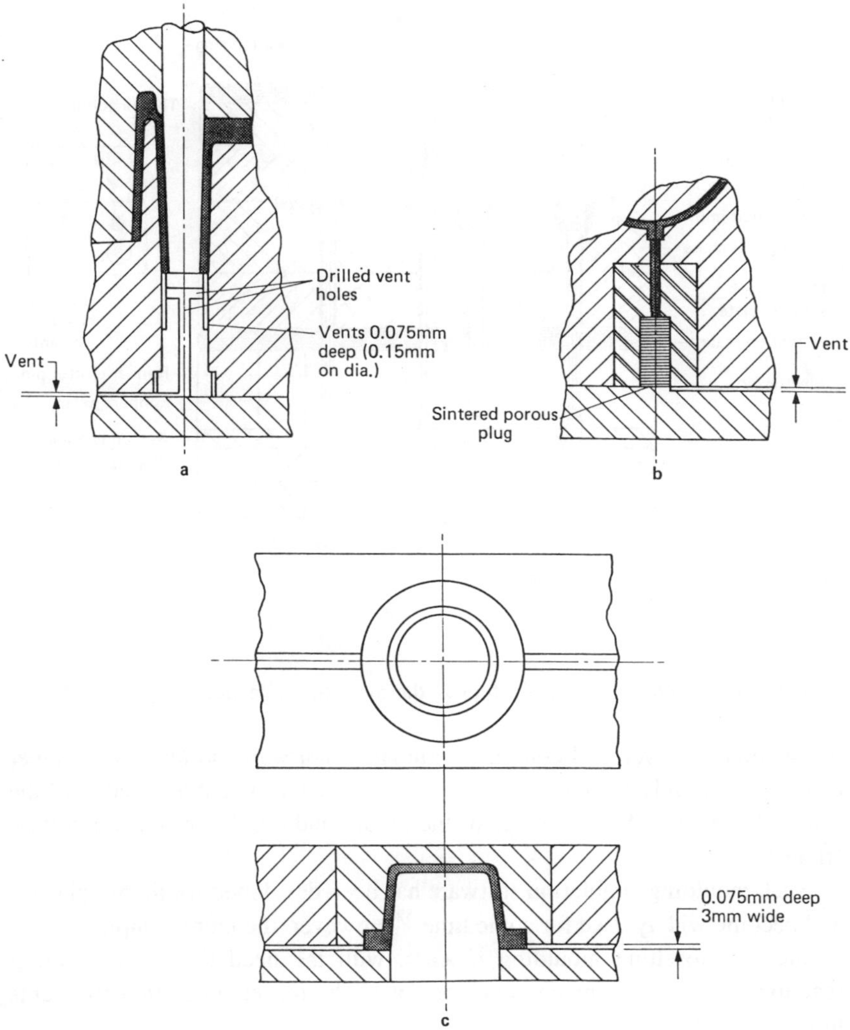

Figure 6.12 Venting approaches. (a) Boss venting, (b) sintered porous plug, (c) die face venting.

achieve good filling with avoidance of stresses, flow lines, voids and short mouldings.

6.6 Thermoset moulding process simulation

The cost and time taken to develop new products is always a major issue, and an important aspect of business competitiveness. It is therefore potentially of great benefit if the flow of plastic into an injection mould can be modelled,

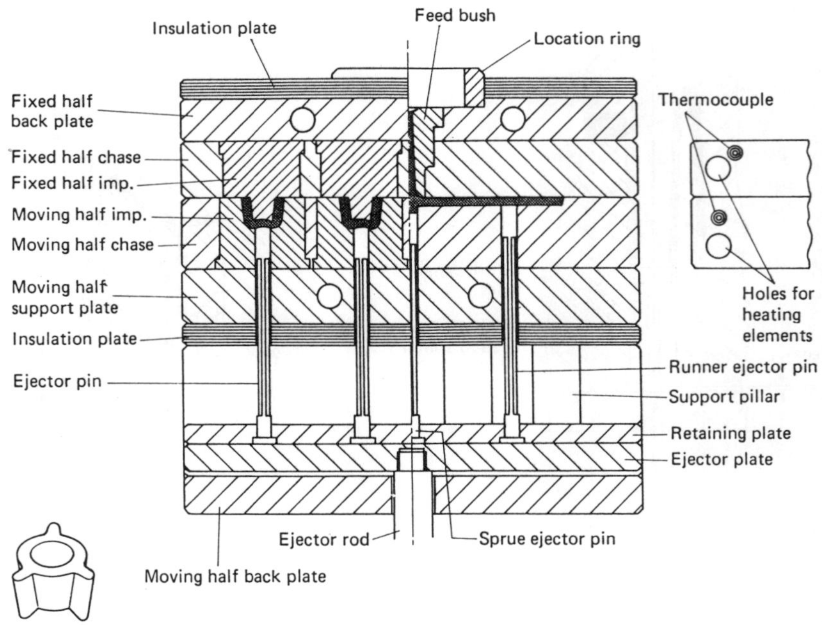

Figure 6.13 Sectional view of a typical thermosetting injection mould.

using suitably developed computer simulation software, so that the designer can design moulds and moulded parts with the certainty that the quality of the finished part, and the behaviour of the mould and moulding process will be 'right first time'.

Such moulding simulation software has been developed for thermoplastics and become widely used for some time.[52] However, the more complex nature of the thermosetting moulding process, with the need to take the curing reaction into account, made development of thermoset simulation programs more difficult.

At the time of writing this work, software for thermoset injection moulding process simulation is now commercially available, and can be expected to be widely used in the future.[53]

The thermal and chemorheological behaviour of thermosetting compounds has already been discussed in Chapter 2, and is covered extensively in the literature. The basis of the models developed for computer simulation is to base the development of the software for the flow analysis on approaches adopted for thermoplastic flow simulation, and then modify the equations to take account of thermoset cure. The steps involved in the flow analysis solution will therefore involve calculation of elemental viscosity, pressure distribution, velocity profiles and temperature distributions as the material flow front expands, correcting at all times for the elemental cure level.

The temperature calculation is based on an equation which takes into account contributions from convection, conduction, viscous heating and curing reaction. Using information held in the materials data files on the curing behaviour of the compound (viscosity, gel time, heat of reaction, etc.) an elemental cure time can be calculated. This value equates to the calculated cooling time in a thermoplastic analysis, and is the time required for the centre of the element to reach the specified degree of cure. The program will then normally go on to calculate the final cure distribution versus time for the complete part and also the final temperature distribution.

In Fig. 6.14 we can see the component wire frame model for a typical analysis, together with three indicative output illustrations. The cavity filling pattern and filling times are shown top right, the display of the required cure times for the different component sections is bottom left (note the long cure time required for the thickest sections), and also we can see a prediction of weld line positions in the bottom right.

The development and successful use of these simulation programs will be heavily dependent on the development and standardization of the appropriate material characterization tests necessary to provide reliable data for the computation. Nonetheless, the number of applications is growing rapidly, as indicated by the selection of references provided.[54–57]

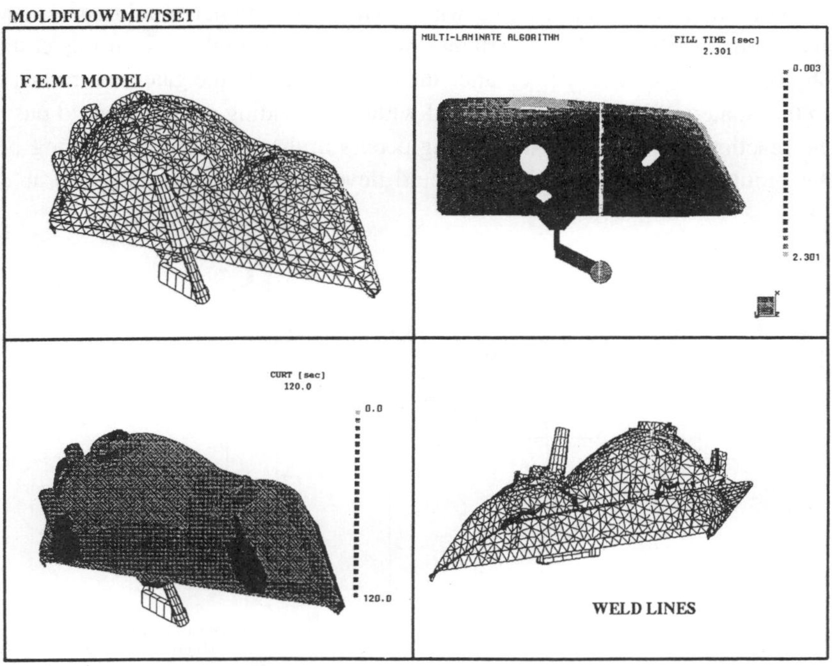

Figure 6.14 Wireframe model and three typical analysis outputs from a thermoset process simulation study made with Molflow MF/TSET system.

6.7 Special injection moulding techniques

6.7.1 Vacuum-assisted moulding

If the air trapped in the moulding cavities is evacuated prior to the entry of thermosetting material, then it is logical to expect that cavity filling can take place much faster, and without the problems of material degradation and gas entrapment often encountered with normal moulding. This potential benefit can also have a bearing on the reduction of problems associated with flow lines, voids and short mouldings.

In principle, vacuum evacuation of the mould offers these potential benefits, and a schematic system to achieve this is shown in Fig. 6.15.

It is important to emphasize that widespread use of this technique is still considered difficult because of the practical considerations of incorporating and maintaining vacuum seals within the heated injection mould. Regardless of these problems, it is likely that this technique will be developed and more widely exploited in future years.

6.7.2 Warm runner moulding

In normal thermosetting injection moulding the sprue and runner sections become cured and irrecoverable waste. The warm runner mould serves an analogous purpose as the cold runner mould for thermoplastics. In this case, obviously, it is necessary to balance the temperatures in the gates and runners so that material fluidity is maintained without providing enough heat to push the reaction rate to a level that curing occurs in these areas during filling of the mould. This calculation and mould development requires considerable

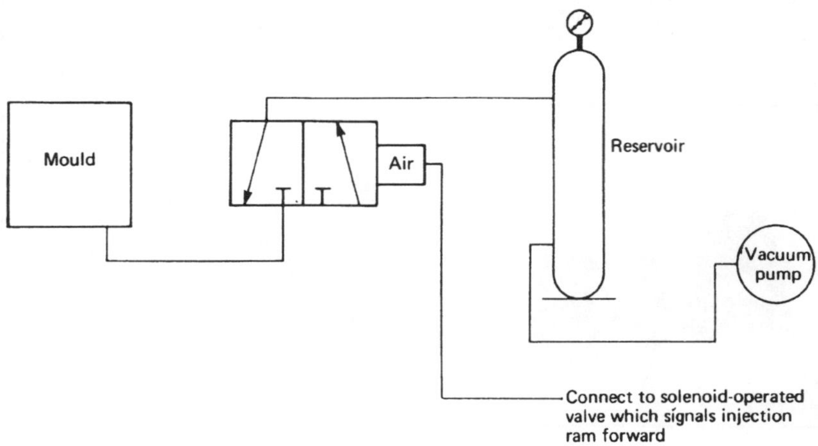

Figure 6.15 Schematic system for vacuum extraction of the mould.

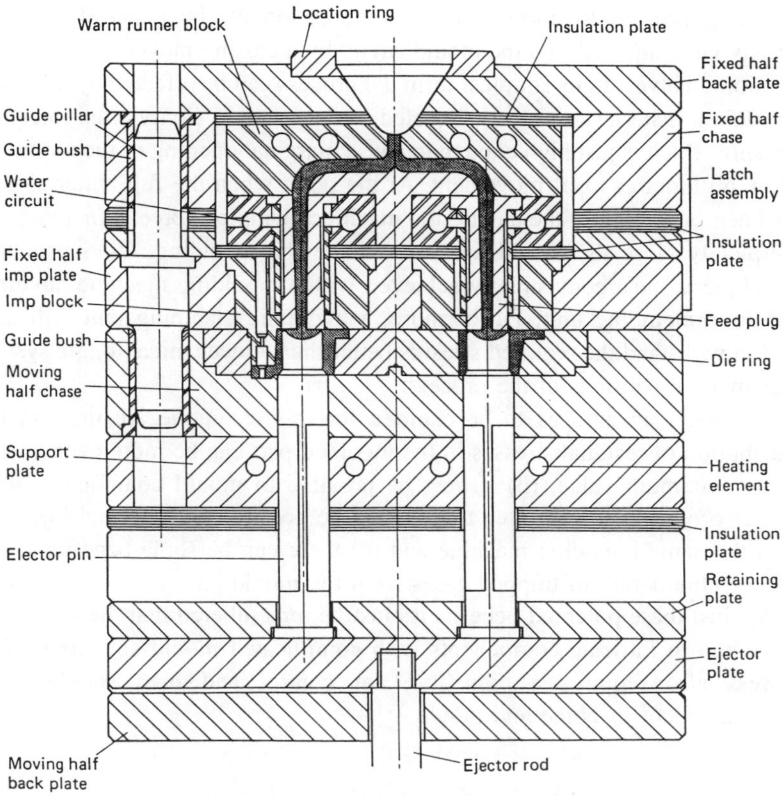

Figure 6.16 Sectional view of a warm runner injection mould to produce the brush box for a starter motor.

skill and experience, and so the technique would not be a recommended choice for organizations new to thermoset moulding, and for products with unsuitable complications. It is likely that computer moulding simulation programs will have an important role to play in assisting this process development. An example of a mould developed for this technique is given in Fig. 6.16.

It needs to be emphasized that warm runner moulding requires close control of the process temperatures together with high regularity of the process cycle to maintain the necessary thermal balance. This in turn means that the mechanical aspects of the mould have to be highly developed to avoid problems of irregular mould function.

6.7.3 Injection compression

The term injection compression moulding describes a variant of the injection moulding process to allow a final compression of the material that has already been injected into the mould cavity by a closing movement of one of the mould halves. In its simplest form the process can be effected by allowing the molten material mass to be injected into the mould cavity at an injection pressure which is just high enough to cause marginal opening of the mould by a few millimetres axial movement. At the point when the full material shot has been injected into the cavity, the machine clamping force is increased to completely close the mould and consolidate the moulding. Obviously this technique is most conveniently achieved with a moulding machine having a hydraulic clamping unit. Furthermore, a hydraulic clamping unit will also tend to withstand the required stresses better than a mechanical toggle system over many repetitions of the process.

The main advantages to be claimed for this technique are that (a) the distribution of residual stresses in the moulded part can be more even and of lower maximum value, (b) it may be possible to mould components with greater projected surface area than would be possible with normal injection from the same moulding machine and (c) there can be some benefits in the ease of elimination of trapped gases from the mould cavity.

Against these potential benefits it must be remembered that the process is not suited to intricate components with complicated mould split lines. The process also requires a high level of control and high standards of maintenance of machine and mould.

6.8 Typical moulding defects

An analysis of defects produced in injection moulds, related to possible causes and remedial actions, is laid out in Table 6.1, while Table 6.2 shows the influence of principal process variables.

Table 6.1 Moulding defects — causes and possible remedies

Defect	Cause	Remedy
Porosity	1. Tool insufficiently vented	(a) Increase or add vents (b) Include breathe sequence
	2. Injection speed too high	Decrease injection speed
	3. Mould temperature too low	Increase mould temperature
	4. Material insufficiently plasticized	Increase barrel temperature
	5. Material too easy flowing	Select stiffer material
	6. Insufficient shot weight	Increase feed volume
Short mouldings	1. Insufficient shot weight	Increase feed volume
	2. Material has too short a flow life	Select longer flowing material
	3. Mould too hot	Reduce mould temperature
	4. Barrel too cold	Increase barrel temperature
	5. If feed setting is on maximum	(a) Increase injection speed (b) Increase barrel temperature (c) Increase back pressure (d) Select easier flowing material (e) Vent mould (f) Increase gate size
Flow marks on moulding surface	Skin curing of injected material:	
	1. Too long a residence time in cavity before completion of fill	Increase injection speed
	2. Material insufficiently plasticized	Increase barrel temperature
	3. Grade of material has insufficient flow life	Selection material with longer flow life
	4. Lack of venting	Increase or add vents
Excessive flash	1. Too much feed	(a) Reduce feed volume (b) Reduce back pressure (c) Reduce injection speed and/or pressure
	2. Gate area too small (multi-cavity moulds)	(a) Balance feed system (b) Increase outer impression gate areas
	3. Material too stiff	Select softer flow
	4. Injection taking place before maximum clamp pressure is applied	Adjust pressure switch
Warped mouldings	1. Too high a pressure gradient within mould cavity	(a) Increase injection speed (b) Reduce hold on pressure (c) Increase gate area (d) Select long-flow material
	2. Cure time too short	Increase cure time
	3. Ejection too fast	Reduce ejection speed
	4. Gate position too asymmetrical	Relocate gate
Blisters on mouldings	1. Component uncured	(a) Increase cure time (b) Increase mould temperature (c) Increase or add vents (d) Increase injection speed (e) Increase barrel temperature
	2. Case hardening of moulding surface	Decrease mould temperature
Cracks in vicinity of gate	1. Over packing	Decrease hold-on pressure and/or reduce hold-on pressure time
	2. Gate too small	Increase gate area

Table 6.2 Influence of principal injection machine variables

Variable	Change	Primary effect	Secondary effect
Barrel temperature	Increase	(a) Increase in injection speed (if previously too low)	Increase in shot weight
		(b) Progressive reduction and variation in injection speed and screw back time (if previously correct)	Decrease in shot weight leading to short mouldings Ultimate effect will be pre-cure in barrel
	Decrease	Decrease in injection speed	Decrease in shot weight
Injection speed	Increase	(a) Increase in porosity (b) Increase in shot weight (c) Reduction in flow marks (d) Decrease in distortion	Reduction in cycle time
	Decrease	(a) Decrease in porosity (b) Decrease in shot weight (c) Possible increase in flow lines (d) Increase in distortion	Increase in cure time
Mould temperature	Increase	(a) Decrease in flow marks	Decrease in cure time
		(b) Increase in surface gloss	Tendency for surface cracking on thick section
	Decrease	(a) Increase in flow marks (b) Increase in porosity	Increase in cure time
Back pressure	Increase	(a) Increase in material temperature	Increase in shot weight
		(b) Increase in screw back time	If too high, pre-cure in barrel
		(c) Higher screw wear	
	Decrease	(a) Decrease in material temperature	Decrease in shot weight
		(b) Decrease in screw back time	Erratic variation in shot weight
Hold-on pressure	Increase	Stress cracking at gate	
	Decrease	Counter pressure from cavity causes 'pumping' action on injection screw	
Screw speed (rpm)	Increase	(a) Increase in material temperature	Decrease in cycle time
		(b) Decrease in re-charging time	Faster injection speed
	Decrease	(a) Decrease in material temperature	Increase in cycle time
		(b) Increase in re-charging time	Possible slower injection speed

CHAPTER 7

Bulk and sheet moulding compounds

7.1 Introduction

This chapter deals with the two increasingly important families of thermosetting polyester moulding materials, the so-called bulk moulding compounds (BMCs), sometimes also called dough moulding compounds (DMCs), and their sister family of sheet moulding compounds (SMCs).

From the early development of these material and associated moulding processes during the late 1940s and early 1950s, major and continuous improvements have been made in formulations, properties and processes. In part, this strong interest stems from the fact that it is possible to develop compounds with excellent and varied performance characteristics at low compound cost by choice and regulation of the proportions of the compound ingredients.[58] Over the last 20 years or so, the composite market worldwide has been one of the most exciting and challenging sectors of the materials industry. Thermosetting polyester moulding compounds have played a significant part in this overall picture. They have also become important materials in the changes that have transformed the humble business of reinforced plastics into the high technology composites industry we see today, in which a large proportion of all fibreglass processed is by the BMC/SMC route.

These changes have stimulated significant market developments in automotive, electrical, general mechanical, building and diverse structural applications. In particular the growth of applications in the automotive industry has been very noteworthy,[59] including use for exterior panels, vehicle substructures such as front-end modules, bumpers, lighting housings and reflectors and other important semi-structural parts. These products have been produced by a variety of moulding processes, including traditional compression techniques but also making increasing use of injection moulding, to gain the benefits of higher levels of productivity and automation that are possible by this route.

This high level of sustained activity over many years has also pushed forward many developments in materials formulation, compounding methods and special moulding techniques. The most important of these advances are highlighted in this chapter, as well as discussion of the basic compression and injection moulding technologies for these materials, which is covered in Chapters 5 and 6.

7.2 Bulk moulding compounds (BMCs)

7.2.1 Basic description and manufacture

Nearly all BMCs are based upon unsaturated polyester resins, typical of which are the polyesterification products obtained by the reaction of propylene glycol with maleic anhydride or isophthalic acid, used in various proportions to achieve the desired resin properties. Such resins are used as solutions in reactive monomers, of which by far the most common is styrene. In addition to polyesters, vinyl ester and phenolic resins are now becoming popular for use in BMC/SMC compounds.

Unlike many of the processes used for the manufacture of composites in which fibres and resin are combined and wetted-out within the process, BMCs are fully compounded products that already contain all the ingredients necessary to mould directly in pressurized heated moulds. As well as the polyester resin, they contain the following ingredients:

- glass fibre, usually between 3 and 25 mm in length;
- an inert particulate filler, chalk, china clay, alumina and many others are used;
- mould-release agent, usually a metallic soap;
- pigments;
- flame-retardant additives, various halogenated organic substances, antimony oxide, alumina;
- catalyst, usually an organic peroxide;
- inhibitors, hydroquinone, catechol, parabenzoquinone;
- low-profile additives, various thermoplastic polymers.

Glass fibre has the obvious effect of strengthening the material,[32] most noticeably improving stiffness and impact strength. Fillers both cheapen the composition and control its flow characteristics, and the catalyst provides the means of initiating polymerization during the moulding process. A mould-release agent is essential to aid extraction of the mouldings.

The curing process is effected by the reactive monomer cross-linking with the polyester resin to produce an infusible solid. This reaction takes place under the influence of the catalyst by a free radical mechanism, which is very rapid. In the case of BMCs, catalysts are chosen that will allow a good storage life for the composition at normal temperatures, but become active at elevated temperatures in the moulding operation. Thus, BMCs are converted to moulded articles in heated moulds held closed under pressure.

Dilatometric measurements with conventional resins have shown that volumetric shrinkage during curing is approximately 7%. However, a very important feature of BMC materials technology is the opportunity to formulate compounds having little or no shrinkage during the curing (moulding) process. These compounds are often referred to as 'low profile' or 'zero shrinkage'.

The mechanism of shrinkage control is based on the incorporation of powdered thermoplastic additives into the formulation. Some materials, such as polystyrene and polyethylene, are non-polar and non-compatible with the resin system, and when added in quantities up to 5–10% provide some shrinkage compensation, with improvements in surface finish and pigmentability. The more effective additives, such as the polyvinyl acetate, polymethyl methacrylate, polyvinyl chloride and cellulose acetate butyrate, are to some extent soluble in the uncured polyester, but unsoluble in the cross-linked resin.

The most commonly held explanation for the events that occur in this phase of the reaction proposals is that the thermoplastic material tends to retain some of the unreacted styrene and non-polymerized polyester resin. As the temperature increases, the higher thermal expansion of the thermoplastic and the vapour pressure of the attached styrene compensate for the polymerization shrinkage of cross-linking polyester/styrene resin. Finally the styrene and polyester residues in the thermoplastic phase polymerize and shrink, causing the formation of voids within the thermoplastic phase.[36,60] As cross-linking proceeds, and very little unreacted styrene remains, strain develops in the system, which increases to the point that stress microcracks propogate through the weak thermoplastic phase to relieve the strain.

A more comprehensive listing of shrinkage control addititives is given in subsection 4.1.6, while the different thermal expansion characteristics of some suitable thermoplastics are shown in Fig. 7.1. The importance of these additives to polyester compounds should not be underestimated, since without their use there would be severe limitations in the processability of polyester compounds by hot press compression or injection moulding.

The pairing of the structures and molecular weights of the unsaturated polyester resin and the shrinkage control addititive is very important for optimum shrinkage control, the additive often being dissolved in styrene to allow good mixing and dispersion into the moulding compound.

The study of shrinkage control mechanisms is a very complex subject, since the low profile effect is acknowledged to involve not only the thermoplastic phase and phase separation during curing, but also the influence of fillers. For example, when magnesium oxide is incorporated in the formulation, as a chemical thickening agent to modify the stiffness of the uncured compound, it is observed that the shrinkage control behaviour of the compound can be significantly changed.

In comparison with other thermosetting materials, BMCs are very easy flowing owing to the low viscosity of the resin and, in fact, are already plastic at normal temperatures. Thus, unlike, for example, a phenolic, no heat has to be put into the material before moulding can occur. Further, very little or no

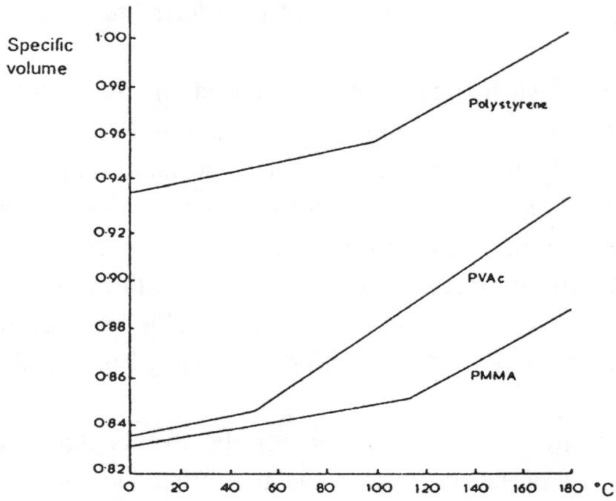

Figure 7.1 Thermal expansion of various thermoplastics as a function of temperature.

curing takes place in the material until the critical temperature of the catalyst is exceeded, so that, although the materials cure very rapidly once this temperature is exceeded, until that time, the flow remains substantially unchanged. These rheological characteristics make it easy to form large, very thin or very complex moulding shapes without the use of excessive pressures. In this respect they differ very greatly from the other thermosetting plastics. It is perhaps in injection moulding that these features become most evident, because quite modest pressures can be used successfully and very narrow and thin runners and gates may be used.

BMCs properties (see Table 7.1) vary somewhat according to the composition arrived at for different application areas. However, they are typically materials having high rigidity and heat resistance, good electrical insulation properties (especially under adverse environmental conditions) and, most of all, having exceptionally low mould shrinkage with unusually good dimensional stability. They have thus been substantially used for technical mouldings requiring high precision, good insulation properties and toughness. Typical applications (see Fig. 7.2) are insulation for electrical distribution equipment, automotive and domestic electrical products, power tool cases and insulation, electric and gas meter boxes, and many applications replacing metal components where the requirements are for rigidity, strength and accuracy.

7.2.2 Comparison of moulding processes

BMC may be moulded by any of the compression, transfer or thermoset injection moulding processes.

Table 7.1 Representative properties for three typical BMC families

Property	Units	General purpose	Low shrinkage	Self-extinguishing
Density	g/cm^3	1.9	1.75–1.85	1.80–1.85
Mould shrinkage (%)	%	0.2–0.3	0–0.05	0.2
Tensile strength	MPa	35–50	35–50	35–50
Flexural strength	MPa	60–110	55–90	60–110
Flexural modulus	GPa	7–10	7–10	7–10
Notched impact	kJ/m^2	16–40	20–35	17–38
Deflection temperature, 1.8 N/mm^2	°C	> 200	> 200	> 200
Surface resistivity	Ω cm	10^{12}	10^{12}	10^{12}
Coeff of thermal expansion	10^{-5} K^{-1}	2.0–3.3	0.8–1.2	1.0–2.5
Fire retardancy	Category	None	None	None
Oxygen index		—	—	33

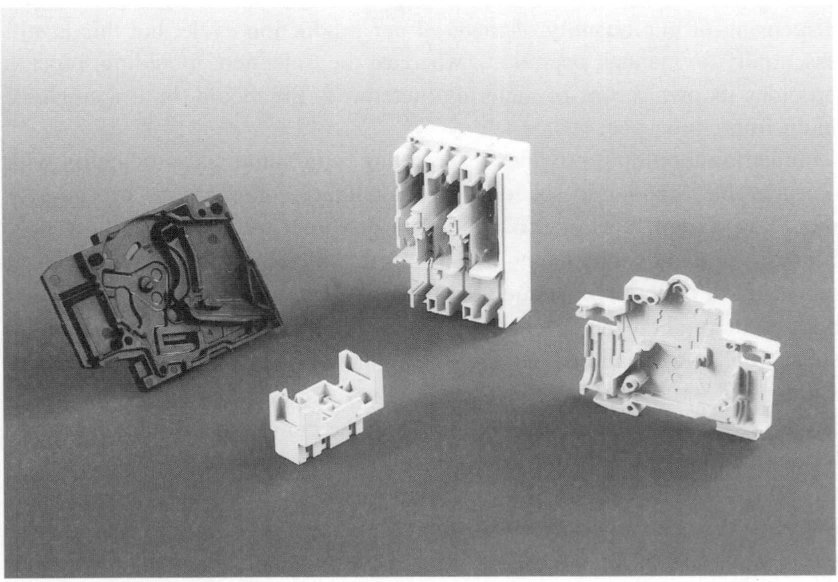

Figure 7.2 A range of typical BMC applications.

Compression moulding is the oldest and most commonly used process for the conversion of BMC to moulded articles. It is the simplest moulding method, and consists of forming the material between the heated metal punch and die, the closing and opening of which is performed in a hydraulic press. There are few components that cannot be made in this way, and the cost of the machinery and tooling is the lowest of the three processes. However, the rather intractable form of the material makes it difficult and tedious to weigh out and then dispense quantities to a multi-cavity mould, and in such instances transfer or injection moulding are preferred, because they avoid this

Table 7.2 Comparison of typical production rates for the different processes

Cycle element	Small moulding (thin section/ 10g weight), 12-impression mould			Large moulding (thick section/ 1000g weight), single impression mould		
	Compression	Transfer	Injection	Compression	Transfer	Injection
Cure time (s)	30	30	25	45	40	30
Press open time (s)	50	15	10	15	15	10
Total cycle time (s)	80	45	35	60	55	40
Saving as a % of compression cycle time	—	44	56	—	8	33

difficulty and reduce the production cycle time and labour costs. Transfer moulding of multi-cavity moulds requires only the weighing out and dispensing of one quantity of material per production cycle, but this is still essentially a manual operation, whereas the injection moulding process provides its own means of metering material to the mould, be it a single or multi-impression one.

Injection moulding provides a means of fully automatic production with considerable economic benefits over either compression or transfer moulding. Faster rates of production are possible (see Table 7.2), not only because it avoids the delay arising from weighing out and dispensing quantities of material to the mould, but also because the material is preheated both by thermal conduction in the barrel of the injection unit, and also, more substantially, by frictional heating as the material passes through the runner system and the gate. As a consequence, the material often enters the mould having reached the mould temperature, and the cure time will therefore only be that necessary for the inherently fast cross-linking reaction to take place. On the other hand, cure times in the compression moulding process are longer owing to the need, not only to carry out the curing reaction, but also to raise the temperature of the material to that of the mould. The latter time can be substantially more than the time to complete the cross-linking reaction for very large mouldings.

In view of this, it is not surprising that cure times for the injection moulding process are much less influenced by the moulding thickness, and very thick mouldings may require very little more cure time than thin ones. The extent to which this is true will depend on several factors, the chief of which is the cross-sectional size of the gate; clearly very small gates will cause a lot of frictional heat to be developed to the benefit of faster cure rates, whereas a mould with a large direct-sprue feed will only heat up the material to a minor degree, and cure rates will therefore be longer in such instances. It follows that faster curing compounds can be formulated for injection moulding.

Metal inserts can be easily incorporated into BMC mouldings during their formation. This is simple and straightforward when the mould is arranged horizontally, as in compression or transfer moulding. Injection moulding, where this is not the case, is much less suited to incorporation of inserts during the moulding process, although it can be done at the expense of slowing down the production rate. Insertion into the moulding can take place after the moulding operation by the use of 'push-in inserts' or inserts with special threads, and this is the usual practice if injection moulding is chosen.

Another difference between the processes is the degree to which flash is formed and the problem of its subsequent removal. Semi-positive moulds have to be used in compression moulding, and this means the loss of some material as flash. As this is reinforced with fibre, it is not as easy to remove as it is with other thermosetting materials, and de-flashing may therefore often have to be done by hand. This is both expensive and less satisfactory, because the finish line may be visually objectionable. Flash is very much less in the injection moulding process, because the material is injected into a mould that is already closed and where the two halves are close fitting. What flash is formed (and there will be some) is sufficiently thin to break away easily, and its thickness is such as not to leave a noticeable scar. Thus, the quality of mouldings made in this way can be very significantly better, particularly in those applications where there are many cored-out holes.

As BMC is a fibre-reinforced material, orientation can be very important. The central loading of a charge of material in a compression mould leads substantially to radial orientation as the material flows out to the extremities of the cavity. The effect of this on the anisotropy of the moulding can be detected, and may be significant in some instances, but generally the effect is not large enough to be very important. Weld lines, however, may be formed where the flow of the material in the mould is obstructed by a core pin, or the shape of the component is such that two moving fronts of material have to join up and in this instance the local strength may be very much reduced. This effect may be avoided or diminished by changing the way that the charge of material is loaded in the tool.

Transfer and injection moulding are likely to give stronger orientational effects because the filling of the cavity in these processes is from one fixed point, the gate. Flow distances are longer, and consequently fibre orientation and weld lines may be more marked and more significant. This has to be taken into account in the design of the part and of the mould, particularly with respect to placement of the gate and its geometry. However, there are ways of minimizing these effects, such as injection compression, and these are discussed in more detail later in this chapter.

There may be a significant difference in quality between parts made by the three alternative processes. On the whole, compression mouldings will provide the strongest components, because passing the reinforced material through the runner system and gate of a transfer or injection mould causes damage to the structure of the fibre and consequent loss of strength. This problem may be diminished by various techniques discussed later in this chapter, but there will always be some difference between the processes that will cause compression moulding to be chosen as the preferred process when maximum strength is important.

Another aspect of quality that divides the processes is surface finish. BMCs are capable of giving mouldings with very smooth and glossy surfaces. However, it is rarely possible to achieve a consistently good and attractive finish by compression moulding, because handling of the material may transfer dirt from hands and working surfaces to the material and because the material is so fast-curing that some minor amount of pre-cure nearly always takes place before the mould can be closed. Further, it is not always possible to use a mould that is positive enough to pressurize the material adequately in a way that is essential for the very best surface finish. Injection moulding differs significantly in this respect, in that the material can be transferred very rapidly from the barrel of the machine into the cavity, and thus pre-cure can be totally avoided. Injection into a closed mould and subsequent very fast cure of the material also helps to achieve the very best finish on the moulding.

There are clearly many factors, both technical and economic, that influence the choice of process for moulding BMC, and these must be weighed carefully before a decision is made. There is a considerable move towards injection moulding because of the significant economic advantages of faster and automatic production and the significantly better visual quality of the parts made.

7.2.3 Compression moulding

Compression moulding of BMC differs from that for other thermoset materials, such as phenolics, ureas and melamines, in that the material is very much easier flowing, involves no gaseous products, such as water or formaldehyde, and cures very rapidly.

A prime requirement for a compression press for BMC is that it should be fast-closing, otherwise the material will pre-cure before the mould is closed. Most modern compression moulding equipment of a standard type will provide more than enough pressure to mould BMC, which, on average, requires only a half or a quarter of the pressure necessary for conventional thermosetting compounds. Whatever pressure is provided by the press, it should be capable of variation, so that the operating pressure may be reduced where this is desired.

Closing speed needs to be fast so that, whatever the size of the press, the closing of the mould can be completed within 10 seconds at the fastest speed.

There should be provision for a final slow-closing stage, which should be controllable with respect to the point at which it comes into operation and the actual speed of closure.

7.2.3.1 Moulding conditions

BMCs are normally moulded at temperatures of 120–160°C, and most often between 140 and 150°C. The higher temperatures are clearly preferred to achieve the fastest production rate, but it is often beneficial to use lower temperatures to achieve the best surface finish, particularly where a particularly slow close is necessary.

It is usual to maintain the same temperature on both punch and die, but it is sometimes a useful practice to maintain a difference between the temperatures of the two halves of the mould, to encourage the moulding to stay on one or other half of the tool when the mould opens. When this is required, the lower temperature (by 5–15°C) is maintained on that half of the tool where the moulding is required to stay.

As BMC flows very easily, even at room temperature, very low pressures are sufficient and often preferred, 35 bar being usually adequate, and even half this pressure is sufficient for the particularly easy flowing grades and simple mouldings. Pressures higher than 35 bar may be required for complex mouldings with bosses or deep blind cavities, and the design of the mould will also be an important consideration. For example, open-flash moulds require less pressure than semi-positive moulds, and very positive moulds may require several times the minimum. In general, 35–70 bar is sufficient for most mouldings, but 150 bar may sometimes be necessary with very positive moulds where the very best possible surface finish is desired.

The way that the charge of BMC is placed in the mould is important. The material should first be consolidated by hand and then, in most instances, placed centrally in the mould which is then closed without delay. There will be occasions when it will be preferable to place the charge differently to avoid an undesirable flow pattern, for example around a core pin, or to assist filling of a boss or rib.

Generally, it is most undesirable to load more than one piece of BMC into a mould as the joining of these when they flow together will form a weak weld line (see Fig. 7.3).

The fast-curing characteristics of BMCs make it essential to close the mould as rapidly as possible after placement of the material. Equally, it follows that procedures such as dwelling or breathing, which are commonplace for other thermosetting materials, are not desirable because some pre-cure is almost inevitable.

Cure times are very short with all grades of BMC, but variations can occur; for example, black materials have been observed to be slower curing

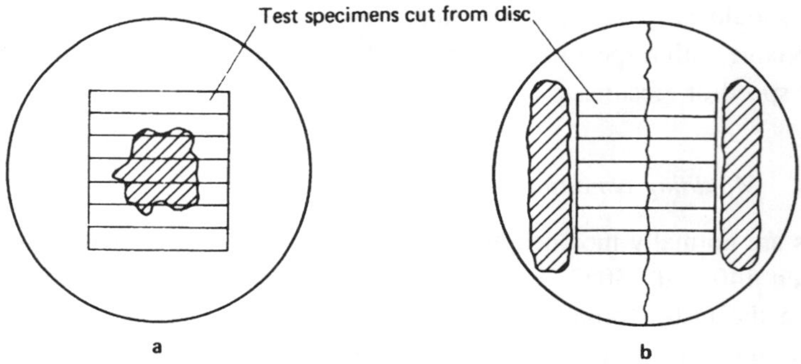

Figure 7.3 Formation of weld lines. (a) Single charge of BMC, no weld lines, flexural strength 70 MPa. (b) Split charge of BMC, weld lines, flexural strength 30 MPa.

than the standard product, probably due to some cure inhibition caused by the pigment. Figure 7.4a shows examples of cure times at different temperatures and thicknesses of moulding, and Fig. 7.4b shows an equivalent set of data for the same material pigmented black.

In practice, it is not possible to over-cure these materials (that is to diminish the quality of the moulding by extending the cure time) and under-cure is almost always self-evident. Unlike conventional thermosetting material, the transition from the state of obvious under-cure (i.e. the material has not solidified) to one of satisfactory cure is very rapid. At an intermediate stage, the BMC may have solidified, but be blistered or have a surface

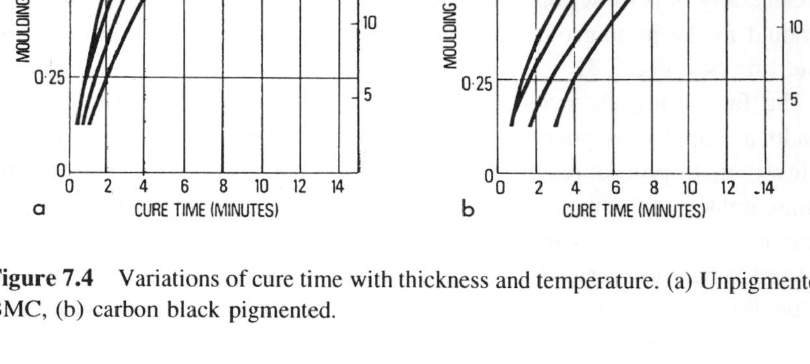

Figure 7.4 Variations of cure time with thickness and temperature. (a) Unpigmented BMC, (b) carbon black pigmented.

entirely lacking in any gloss, but once these deficiencies have disappeared and the moulding appears visually satisfactory, then in nearly all instances the full quality and properties of the moulding will have been developed. Because visual examination is so reliable, it is not usual to carry out any more sophisticated cure tests, such as acetone extraction, as is common practice with phenolic mouldings.

The common problems in moulding BMC are listed in Table 7.3, together with their causes and possible remedies.

7.2.3.2 Mould design

Moulds for BMC are usually designed on the same principles as those applicable to conventional thermosetting materials. They should be made with good quality, hardened alloy steels and finished to a high polish on all mould faces. Chromium plating is desirable for long life and easier ejection of the mouldings. The very easy flow of the resin will allow it to fill any joint lines within the cavity, even if these are very carefully matched, and for this reason construction of the mould from several separate pieces must be avoided, unless these pieces are loose, so that flash is cleared from them on each moulding cycle.

The usual design of mould for BMC is semi-positive with vertical or horizontal flash, but open flash moulds may sometimes be used successfully for relatively simple shapes. Fully positive moulds are unsatisfactory, since they are difficult to vent effectively, and small variations in the mould charge will result in mouldings of different thicknesses.

General features of mould construction are not given here, but can be found in Chapter 5. Only those features peculiar to BMC compression moulding are described below.

Vertical flash moulds, exemplified in Fig. 7.5, are much preferred because this provides more positive pressure on the material rather than on the lands, and a moulding of better quality surface finish is thus obtained. It should be noted that pressure pads are provided on this type of mould, to define the moulding thickness and protect the cavities.

Although vertical flash moulds are preferred, their construction for some parts of a complex shape can be both difficult and very expensive, and in such cases a horizontal flash, semi-positive mould will have to be used (see Fig. 7.6).

The essential feature of such a design is the horizontal landing face, and it should be realized that, as BMC cures very rapidly, some material will be trapped between the land faces towards the end of the closing operation, and this will to some extent reduce the effective pressure on the actual moulding; the effect is accentuated by the fibre reinforcement in the BMC. As a consequence, the land width should be less than for conventional moulding

Table 7.3 BMC compression moulding guide to faults, with causes and possible remedies

Defect	Possible causes	Remedies
Poor gloss	(a) Under-cure (time too short and/or temperature too low)	Modify mouilding conditions accordingly
	(b) Moulding pressure too low	Modify moulding conditions accordingly
	(c) Temperature too high	Optimum temperature for best finish is 132°C for colours or 140°C for blacks
Gloss only on one side of mouldings	Uneven mould temperature	Equalize punch and die temperature
Soft mouldings	(a) Under-cure (time too short and/or temperature too low)	Modify moulding conditions accordingly
	(b) Use of BMC that has exceeded its storage life or has lost styrene by exposure to air	Replace by fresh material and store in tightly closed containers
Large blisters	Inadequate cure	Increase curing time
Scorching (especially with types containing sisal fibres)	Moulding temperature too high	Ensure that temperature of 160°C is not exceeded
Local burning on webs and bosses	(a) Mould closed too rapidly	Introduce slow close in moulding cycle
	(b) Material too stiff	Use easier-flowing grade
	(c) Lack of venting	Modify mould accordingly
Internal cracks or voids	(a) Shrinkage of thick sections	Use suitable grade of material
	(b) Insufficient cure	Increase cure time
Internal porosity	(a) Press closed too rapidly	Introduce slow close in moulding cycle
	(b) Material flowing too easily due to:	
	1. Pre-heating	Omit pre-heat
	2. Use of short dwell	Omit dwell
Moulding cracks on ejection	(a) Uneven ejection	Clear flash from ejectors
	(b) Inadequate cure	Increase cure time
	(c) Mould temperature too high	Reduce moulding temperature
	(d) Ejectors wrongly placed	Modify mould accordingly
	(e) Insufficient number of ejector pins	Modify mould accordingly
Excessive flashing down ejector pins	(a) Material flowing too easily	1. Alter mould temperature 2. Reduce moulding pressure 3. Use material of stiffer flow
	(b) Ejector clearances excessive	Fit pins with minimum clearance
Oversize mouldings	(a) Low moulding pressure	Increase pressure
	(b) Material curing before mould is fully closed	Reduce moulding temperature, increase press closing speed or, if dwelling, shorten dwell time

	(c) Mould too positive for easy escape of flash	Increase cavity clearance at flash line
Short or porous mouldings	(a) Low moulding pressure	Increase pressure
	(b) Premature gelation preventing complete flow	Reduce moulding temperature, shorten dwell time or omit pre-heat
	(c) Venting on bosses or webs inadequate or not provided, causing trapping of air	Provide vents at points where shorts occur
	(d) Material too stiff flowing	Use easier flowing type, pre-heat or give a short dwell
	(e) Incorrect location of charge	Try alternative loading positions or preformed shapes
Swollen white fibrous patches	If material has not exceeded its storage life, may be due to pre-cure	Reduce dwell time or omit pre-heat. Try faster closing speed
Poor mould release	(a) Under-cure	Increase temperature or curing time
	(b) Excessive moulding pressure	Reduce pressure
	(c) New tool	Apply mould release agent for first few lifts
	(d) Excess charge in too positive a mould	Reduce charge weight; if unsuccessful, increase cavity clearance
	(c) Undercuts on mould	Inspect mould and rectify

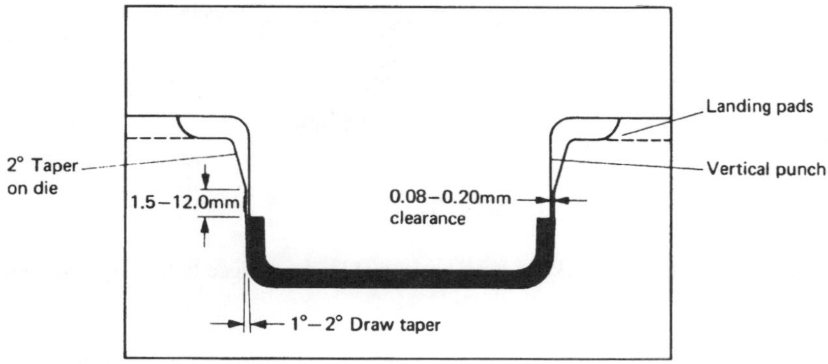

Figure 7.5 Vertical flash semi-positive mould for BMC.

materials in order to reduce flash to a minimum (see Fig. 7.7). The preferred land width depends on the size of the moulding to some extent, but it should be limited to 1.6–5.0 mm. A positive clearance should be provided between the land faces, as it is impossible to reduce the flash to very small thicknesses. This clearance should be 0.05–0.08 mm, and this can be defined as in vertical flash mould designs, by providing pressure pads at the extremities of the bolster. This is particularly necessary to protect the land faces from damage, which would occur if the total moulding pressure were not absorbed by the pressure pads.

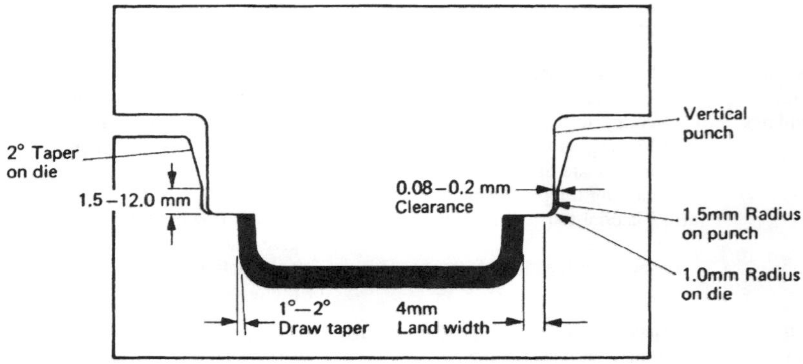

Figure 7.6 Horizontal flash semi-positive mould for BMC.

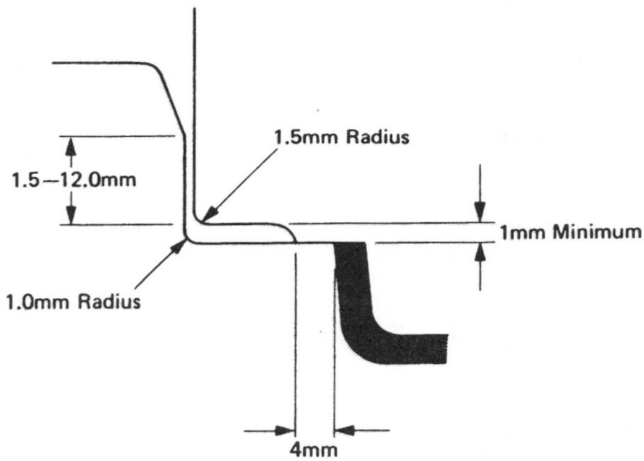

Figure 7.7 Method of backing away horizontal land to reduce to the required width on a horizontal flash mould.

The cavity design (i.e. the design of the mould beyond the impression form) is the most important feature of the total mould design. The purpose of this is twofold: firstly, it should allow air trapped in the mould to escape; secondly, it should allow the escape of excess material. At the same time, the flash line must not allow too easy escape of material, otherwise pressure will be lost in the cavity and the moulding will have a poor finish. These considerations, coupled with the very easy-flowing character of the material, make it necessary to use small and very critically designed flash clearances. The best compromise of all the requirements for the flash design is met with a vertical flash design (Fig. 7.5), which provides for a degree of parallel fit between the punch and die of up to 5 mm in length with a clearance of 0.08–0.18 mm.

Beyond this clearance, the die should carry a 2° taper; the purpose of this is to allow flash that is cured to escape and so avoid blockage of the cavity.

Horizontal flash design is less versatile, but Fig. 7.7 shows that the essential features of flash clearance and cavity taper are similar to those used in vertical flash design. If the horizontal land portion has to be wider than that indicated above, owing to the complex shape of the cavity, then it is important to back away the punch to a horizontal clearance of 1.0–1.5 mm, as shown in the illustration. This will avoid excessive land area, which would absorb much of the moulding pressure to the detriment of moulding quality.

Displacement moulds, such as illustrated in Fig. 7.8, are particularly useful for BMC when making deep draw shapes. In this instance, the cavity should be very positive, because escape of trapped air is provided for by having a vent pin at the top of the mould. These conditions allow good compaction of the material in the mould and a ready avoidance of porosity, together with good surface finish and general moulding quality.

Escape of entrapped air within the cavity is one of the prime considerations of BMC mould design, and this is substantially achieved by proper design of the flash line. However, subsidiary means of air escape are essential on blind recesses of the mould, such as bosses and ribs, and this is done by providing local ejector pins down which entrapped air can escape. Such pins should be designed as shown in Fig. 7.9, and it should be ensured that the pin is a good sliding fit in the hole. A small amount of flash will escape down the pin, but this will be cleared on each stroke of the ejectors, so ensuring the proper venting can occur on the next moulding operation. Open vents must not be used, since BMC will flow through them excessively and will fill the vent hole completely.

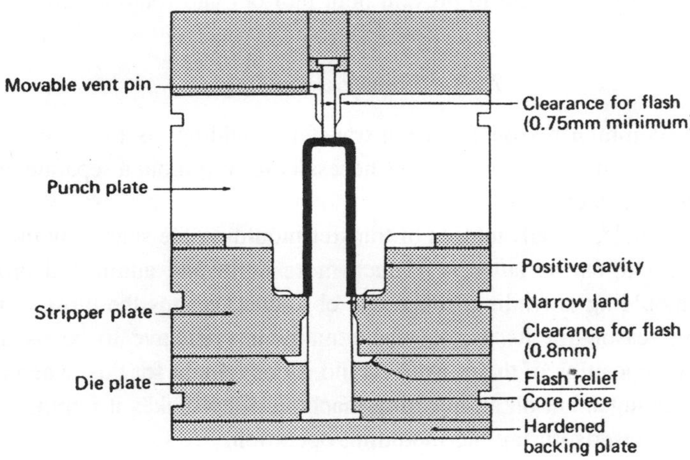

Figure 7.8 Displacement mould.

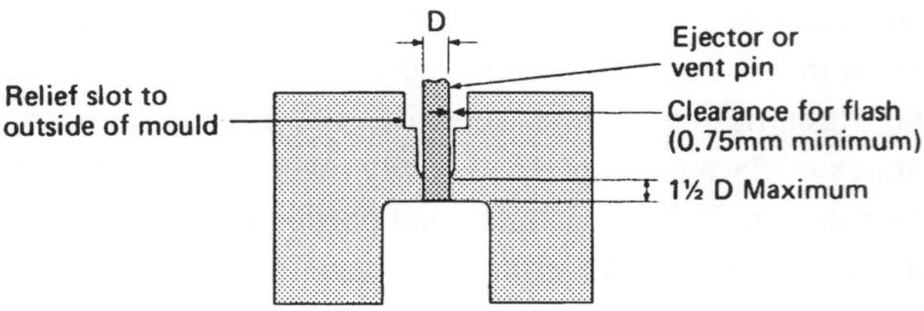

Figure 7.9 Ejector/vent pin design.

All BMCs shrink very little, and some have no mould shrinkage at all. Mould design and construction must take this into account, and makes accurate mould making essential, as even minor roughness or slight back-taper will cause the mouldings to be held in the mould. It is necessary in most cases to provide positive means of ejection of the moulding, and this usually consists of ejector pins that should be evenly distributed along the mould faces having the least taper. These should have the same design as the vent pins illustrated in Fig. 7.9, and indeed the same pins will often serve both purposes.

Side cores may be used quite successfully with BMC, although there is a danger of adjacent weak weld lines, owing to separate flow of the BMC around them. Alternatively, it is often possible to arrange that the side core is brought in after closure of the mould, and in this instance the weld line will be avoided. Moving side cores should be constructed with a minimum clearance, since excessive flashing down this clearance will be troublesome and cause erratic operation.

It is common to mould-in metal inserts, especially where a threaded hole is required, and this presents no special problems. Moulded inserts should have a heavy knurl or undercut to prevent both pull-out and rotation of the insert.

7.2.4 Transfer moulding

The most common reason for using transfer mouldings is to accommodate multi-impression moulds without the necessity to weigh out a separate charge of material for each cavity.

On the whole, the advantages of transfer moulding are shared by injection moulding, and, as the latter is a much more rapid and automated process, transfer moulding is tending to become obsolete. Perhaps the most valuable remaining reason for its use is where many inserts have to be used; the horizontal deposition of the transfer mould is convenient for this, whereas the different arrangement on an injection machine often makes it impractical or costly to fit inserts during the moulding operation.

Of the two types of transfer mould, the integral type (see Fig. 7.10) is the more common, but where the auxiliary ram type is used it is possible,

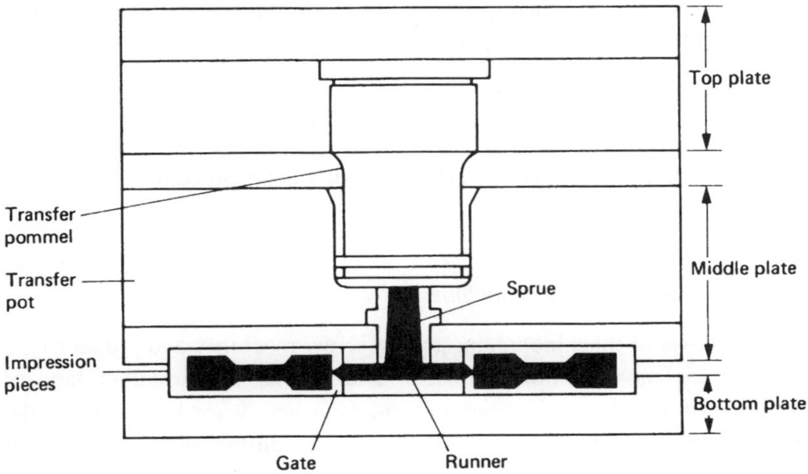

Figure 7.10 Integral pot transfer mould.

and often desirable, to maintain the pot at a considerably lower temperature than the mould itself. This facilitates moulding by keeping the BMC at a relatively higher viscosity, so making it easier to control the flow. The strength of transfer moulded parts is sometimes less than that of equivalent compression moulded ones, owing to some damage of the glass fibre in passing through the runner system and gate, and there may also be disadvantageous variation of strength owing to orientation. The effect of the former is not usually great and can be minimized by using runners and gates of the largest possible size. The effects of orientation should either be taken into account in the design phase so that orientation is either favourable or disturbed by the shape of the cavity at the point where the material first enters the mould.

The geometry of the whole of the runner system will follow similar lines to that for an injection mould, and the reader is referred to the next section.

Moulding pressures are commonly much less than those required for granulated thermosetting moulding materials, and are usually in the region of 140–280 bar. Mould temperatures may be higher than for compression moulding for the auxiliary-ram type of transfer mould where the pot is kept at a temperature substantially below that of the mould. In such instances, mould temperatures as high as 170°C may be used, because of the rapid rate of transfer material into the mould. With integral-pot designs, the pot will be at the mould temperature, and in such cases the same rules apply as for compression moulding temperatures.

Cure rates for transfer moulding are rarely very much faster than for compression moulding at the same temperature, but production rates can be improved because of the facility of feeding a multitude of cavities from one charge of BMC in the transfer pot.

7.2.5 Injection moulding

Injection moulding is fast becoming the predominant process for moulding BMC.

7.2.5.1 Injection moulding machines

Two main types of machine, reciprocating-screw machines and plunger-type machines, are most commonly used. The former are usually standard injection machines having a 'thermoset' injection screw and barrel, where the screw is usually a relatively short feed screw with constant pitch and constant feed depth (or with a small increase in feed depth towards the front). A screw check ring is not always used. If, however, it is specified, the ring and seat must be carefully designed to minimize damage to the glassfibres. The barrel is normally jacketed for fluid heating (Fig. 6.1). This serves suitably for all thermosetting materials, including BMC, but, whereas granulated thermosets can be fed into this design using a gravity-feed hopper, BMC requires a force-feed hopper (see Fig. 7.11).

The force-feed hopper is a simple device consisting essentially of a hydraulically pressurized ram and a cylinder that forces the BMC into the start of the screw. The screw machine has the advantage of being adaptable for other materials (i.e., granular thermosets or thermoplastics), if a different screw and cylinder are substituted, but there are two significant disadvantages. Firstly, the capacity of the injection unit may not be as high as the hydraulic pressure on the ram of the injection unit could allow (taking account of the low viscosity of the BMC). Secondly, rotation of the screw causes some damage to the glass fibre in the BMC, and consequent strength reduction.

The significance of the first point can be eliminated if the machinery manufacturer designs an especially large-diameter injection unit that takes account of the modest injection pressures required by the material. The strength reduction caused by the screw, although partially unavoidable, may be insignificant in comparison with later damage caused by the material passing through the runners and gate of the tool. Much depends upon the type and size of the component being moulded. Small components, for example, may not be very critical for strength, and, in any case, are usually made with very small gates, where there is a high degree of fibre damage and strength loss. In such cases, the minor amounts of damage taking place under the action of the screw becomes quite unimportant.

Large mouldings fed directly from a large sprue are hardly affected by passage of material through such a feed system; in such instances, where the

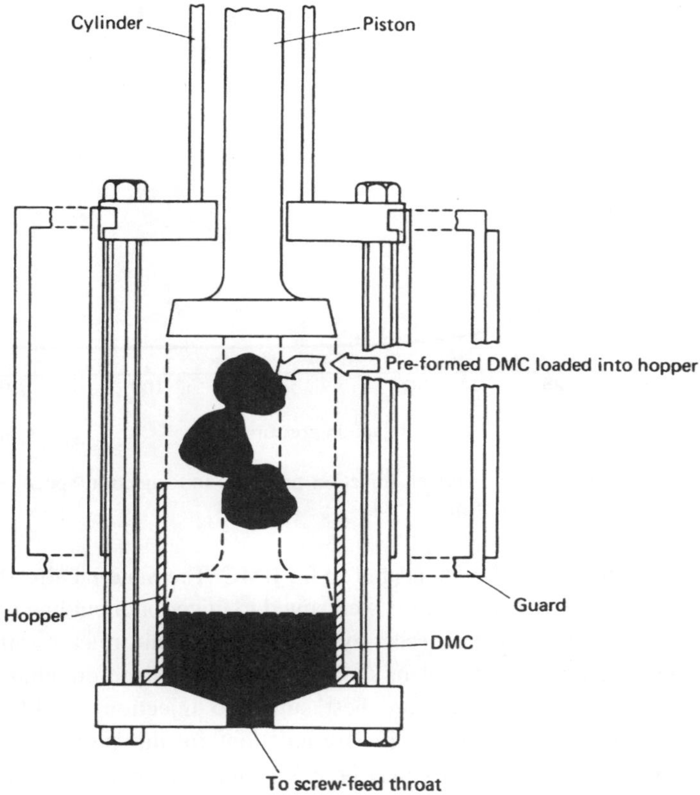

Figure 7.11 BMC force feed hopper.

maximum possible strength is required, any loss of strength from the injection unit becomes important.

The alternative plunger type of machine avoids this difficulty, but the shot is often less homogeneous, and it is not possible to heat the material significantly. Further, there are often problems with wear on the plunger seals.

Typical thermoset injection machines are provided with up to 2000 bar of available injection pressure; such high pressures are necessary because of the inherently high viscosity of the most thermosetting plastics melts. BMC, however, has much lower viscosity and injection pressures of the order of 250–500 bar are normally sufficient. It is thus possible to injection mould much larger moulding weights for the same size of machine, this is clamp pressure and platen size.

7.2.5.2 Material requirements for injection moulding

Conventional thermosetting materials have to be substantially modified to

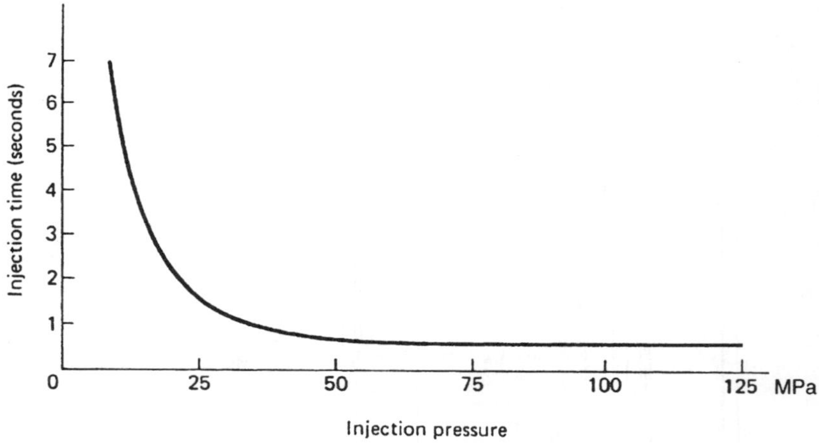

Figure 7.12 Relationship between injection pressure and injection speed, showing the easy flowing nature of BMC.

suit injection moulding, but this is not so for BMC. The material already has a sufficiently low viscosity, which is well suited to injection moulding (see Fig. 7.12), but modifications are necessary to improve mould release, minimize fibre degradation (and consequent strength loss) and reduce orientation. Not all grades of BMC are therefore best suited to injection moulding, and manufacturers usually make specific formulations for this purpose.[61]

Various glass fibre contents may be employed, but on average there is a diminishing proportional strength benefit from increasing glass content, owing to the fact that the glass fibre impedes the flow of the material, and thus the higher the glass content the more damage is done to the fibres, and the greater will be the proportional fall in strength. Typically, most injection grades contain between 12 and 20% of glass fibre, usually of 6 mm length.

7.2.5.3 Moulding conditions

Typical injection moulding conditions for BMC are given in Table 7.4.

As BMCs are already plastic at room temperature, it is not essential to heat the material in the barrel of the injection machine before injecting it into the mould. Nevertheless, some heating at this point will reduce the viscosity and so minimize fibre damage, and it is common, therefore, to use barrel temperatures up to 60°C on screw-injection machines. This is not practical on plunger machines, owing to the poor thermal conductivity of the material and the very short time available for transfer of heat into the mass of material.

Temperatures above 60°C are not normally used because of the danger of the material setting-off in the barrel, but below this temperature the stability of the material is very good, and most compounds can be held in the barrel

Table 7.4 Recommended injection moulding conditions

Mould temperature	140–160 °C
Barrel temperature	20–60 °C
Screw speed*	20–50 rpm
Back-pressure on screw*	15–60 bar
Injection pressure	250–800 bar
Injection stroke (time)†	1–8 s

* Lowest values minimize fibre breakdown.
† Fastest speed gives the best surface finish

under these conditions for a few hours at least without danger of pre-cure or increased viscosity.

Mould temperatures are similar to those used for compression moulding, usually 140–160°C, but higher temperatures, up to 175°C, are often possible, because the rapid transfer of material into the mould makes it possible to use these high temperatures without risk of pre-cure.

The screw speed (see Fig. 7.13) and back-pressure on the injection unit should be kept as low as is consistent with obtaining even shot weights, as both of these parameters have an adverse effect on fibre damage in screw machines. Injection pressures are commonly 250–800 bar, depending on the moulding shape and the geometry of the feed system.

The injection time may be very short indeed, sometimes less than one second. Such fast speeds will usually give the best surface finish and the fastest rate of cure, and are usually preferred. The finish is so favoured because all possibility of pre-cure is totally avoided in such a short time, and the cure rate is much enhanced because shear heating is considerable at such high flow rates. Care must be exercised in the provision and placement of local venting, to avoid surface porosity in remote areas of the moulding.

Unlike compression moulding, cure times for injection moulded BMC parts are not substantially dependent upon the moulding thickness, because the usual expectation is that the material will have been pre-heated to the mould temperature by the time it enters the cavity, firstly by thermal conduction in the barrel and the runner system and secondly by the development of frictional heating as the material passes through the heat

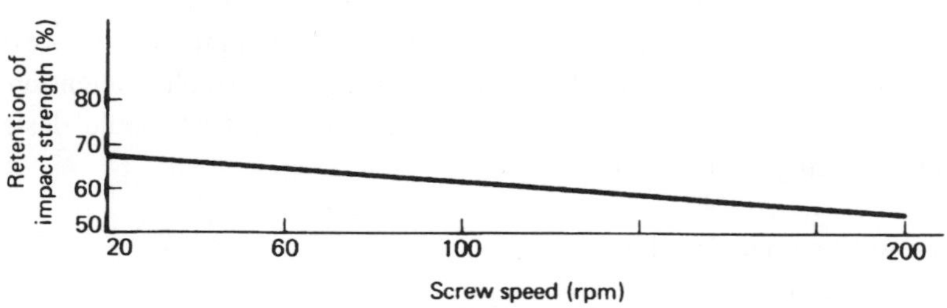

Figure 7.13 Effect of screw speed on impact strength of BMC.

Table 7.5 Relationship of section thickness to cure time for a range of typical components

Moulding	General thickness (mm)	Compression cure time (s)	Injection cure time (s)
Switchplate	3.5	35	25
Electrical plug	4	40	20
Pan handle	20	70	45
Connector strip	6	30	15
Instrument housing	3	40	40
Electrical insulator	30	270	60
Lamp holder	20	75	30

system. This effect, coupled with high mould temperatures, can give very fast cure rates, which may be as short as 15 s in favourable circumstances. Table 7.5 gives examples of cure times with a range of different mouldings and compares these with times that would be necessary with compression moulding. The advantages on very thin mouldings may be only modest, say up to one-third better, but the gain becomes much greater as the moulding thickness increases.

Sometimes cure times are effectively extended beyond those necessary for the thickness of the moulding, because the sprue is substantially thicker than the part being made and benefits very little from shear heating. Due care has to be taken of this, therefore, and it is best to eliminate this factor where possible by using shortened sprues and extended nozzles.

BMC material may be left in the hopper and the barrel for long periods of time at room temperature without risk of the material solidifying. However, when the machine is shut down overnight or for other short periods of time, the bulk of the material should be purged from the barrel with the nozzle removed. The machine should be thoroughly cleaned if left for long periods of time.

A guide to the solution of the common moulding problems is given in Table 7.6.

7.2.5.4 Preservation of strength

The injection moulding process inevitably causes some reduction in the strength of parts, as compared with those moulded by compression. However, the extent of this loss of strength can be relatively very small if attention is paid to using the best techniques of moulding and the best mould design. Factors affecting strength are

- screw or plunger machine
- screw speed
- back-pressure

Table 7.6 Guide to BMC injection moulding faults, with causes and possible remedies

Defect	Possible causes	Remedies
Poor gloss	(a) Lack of pressure in cavity	1. Increase injection pressure 2. Maintain hold of pressure 3. Back away lands to increase clamp pressure 4. Faster injection
	(b) Lack of shot weight or poor transfer of shot	Make injection more positive by use of non-return valve on screw
	(c) Slow rate of cure	Increase mould temperature
Short moulding	(a) Insufficient shot weight	1. Increase shot setting 2. Use non-return valve on screw
	(b) Poor packing of barrel	1. Increase back-pressure on screw 2. Increase screw speed 3. Reduce barrel temperature
	(c) Premature cure	1. Reduce mould temperature 2. Increase injection speed
	(d) Restriction of flow into mould	Increase gate size
Porosity on surface and extremities	Air trapping	1. Provide local venting on mating face and/or with vent pins 2. Reduce injection speed 3. Apply vacuum to mould immediately before injection
Excessive flash	(a) Insufficient clamping pressure	1. Increase clamp 2. Back away landing face 3. Reduce injection pressure
	(b) Excessive shot	Reduce shot setting
Seizure of screw	(a) Pre-cure of material	Reduce barrel temperature to 60 °C or less
	(b) Heating of nozzle from mould	1. Use sprue break procedure 2. Employ separate control of temperature on nozzle (essential for long nozzles)
Local burning	Dieseling of compressed gases	Treat as for porosity to remove entrapped air
Sprue sticking	(a) Insufficient taper	1. Increase taper to at least 3° 2. Reduce sprue length by using extended nozzle in barrel
	(b) Over pressurized	Reduce injection pressure
	(c) Under-cure	1. Increase cure 2. Ensure heating at sprue is adequate 3. Increase mould temperature
Poor strength	(a) Weld lines caused by core pins	1. Relocate gate 2. Use injection–compression technique
	(b) Excessive fibre damage	1. Enlarge gate 2. Ensure well radiused smooth runner system 3. Reduce injection speed 4. Use non-return valve on screw 5. Reduce screw speed and back-pressure

- injection speed
- nozzle size
- sprue and runner size and length
- geometry of runner system
- gate size and location
- moulding geometry

It has already been mentioned that plunger machines cause less damage to glass fibre during material transportation to the injection nozzle. Practice, however, shows the major damage to the glass fibre and hence loss of strength is undoubtedly due to the passage of the material through the runner system. The size of the nozzle of the barrel, the cross-section and length of the sprue and runner, and the size of the gate can all markedly influence fibre damage. For maximum strength, the cross-sections of the whole runner system should be as large as practical, and the length of the feed system should be restricted. Less obvious is the need to provide generous radii wherever the feed system has to change direction, that is from sprue to main runner, from main runner to sub-runner and sub-runner to gate.

The geometry of the part can have a profound influence on strength. For example, core pins or irregular shapes in the cavity that oppose the easy flow of material into the mould will cause additional damage to the glass fibre. Additionally, the shape of the part will influence the degree of fibre orientation and thus affect the average strength obtained in the component. Care therefore needs to be taken to locate the gate in the most advantageous position relative to the shape of the part; in most instances this will be on the heavier section of the moulding.

The maximum strength will always be obtained if a direct sprue feed can be used, as this will create the minimum of disturbance and damage to the glass fibre. Injection speed can have a significant effect on impact strength; the higher the speed, the more damage will occur. Consequently, injection speeds should be kept low if maximum strength is required, although this will be at variance with achieving the best surface finish, which is favoured by fast injection speeds.

Some orientation of glass fibre during the injection moulding process is inevitable, and where this occurs there will be variation in both the strength and the shrinkage between different directions. The differences in shrinkage may be minimal with the so-called 'low-profile' grades of BMC, which have very little or no shrinkage. Location of the gate near to a projection in the mould will prevent the material jetting into the cavity and cause it to be broken up, so completely disturbing orientation. This can be very effective, although it may cause some damage to the glass fibre. Alternatively, and often preferably, the shape of the gate may be altered. Tab gates of small width may cause considerable orientation, especially on thin and essentially

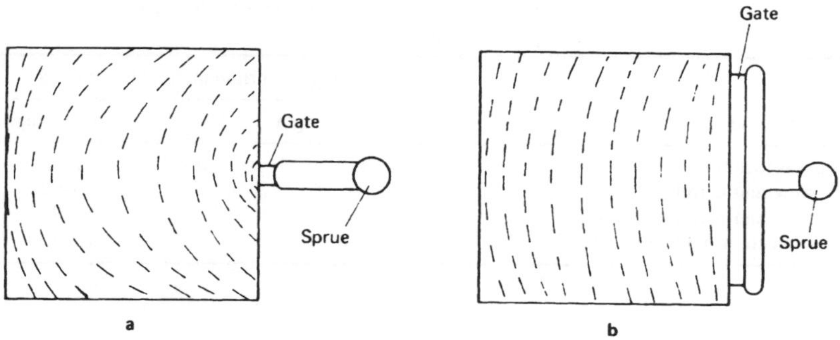

Figure 7.14 Fibre orientation in a panel moulding. (a) Tab gate — strong orientation of fibres results in a very anisotropic moulding. (b) Flash gate — broad feed through gate produces much less orientation.

flat components. The alternative use of a very wide, thin flash gate will substantially avoid this problem by feeding the material over a larger width of the moulding (see Fig. 7.14). This is also a better choice of gate from the point of view of finishing.

Weld lines, another form of orientation that can cause reduction of strength, are formed where the flow from the point of injection is divided by, for example, a large core pin. As the material joins up on the other side of the pin, it does so without benefit of fibres bridging the join, and there is therefore a local weakness. This may be avoided by relocating the gate, or by using a broader gate, or by a technique known as injection–compression moulding.

This technique, which is also discussed in section 6.7, involves injecting the material into a mould that is still fractionally open, and then, when all the material is injected, closing the mould. Under these conditions, the core pin does not completely obstruct the flow of the BMC, and a weld line is not formed. This technique can be used wherever orientation has to be avoided from whatever cause, as the compression stroke completely disturbs any fibre alignment.

In summary, if good techniques and the more suitable mould designs are used, it is generally found that the overall strength of an injection moulded part is only marginally less than a compression moulded one. Table 7.7 gives some comparative data on this topic, based on practical trials on a range of components with different gate sizes and geometries. Most usually, it is the impact strength that is more affected than flexural or tensile strengths, because this property is very sensitive to the structure of the glass fibres.

7.2.5.5 Mould shrinkage and stability

Injection moulding of granulated thermosets tends to produce parts that are

Table 7.7 An indication of retained strength at different stages of the injection moulding process

Stage of process	Percentage or original strength retained	
	Flexural strength	Impact strength
Original material	100	100
ex screw and barrel	—	70–90
ex screw and barrel and nozzle	—	60–80
Final injection moulded part	60–85	40–70

less stable dimensionally than compression moulded ones, and the reasons for this are built-in stress and orientation. Injection moulded BMC parts, however, are not so disadvantaged, because the very easy flow of the material avoids putting stress into the moulding. Further, orientation which does occur with BMC, only becomes serious when it affects the directional shrinkage. As earlier described, low-profile types of BMC avoid this problem by having minimal or nil shrinkage. Where general purpose types have to be used, orientation may present a slight problem of distortion, and in such instances it is best to eliminate this by the use of wide flash gates, which minimize orientation.

In the best practice, it is found that there is very little, if any, difference in the quality of injection moulded and compression moulded BMC parts in these respects; the same high order of accuracy and stability on ageing is observed.

7.2.5.6 Mould design

The design of moulds for the injection moulding of BMC follows many of the same basic principles that are used for other injection moulds (see Chapter 6), and the features peculiar to BMC in compression mould designs previously described are, in many respects, equally appropriate in injection moulds.

Mention has already been made of the importance of the feed-system design, and it is chiefly in this respect that mould design for BMC differs from that for other thermosettting materials. Figure 7.15 highlights the distinguishing features of a BMC injection mould.

BMC flows very easily and it is therefore possible to use a feed system and gate of very much smaller size than could be tolerated for other thermosetting plastics materials. However, as already explained, the largest size of feed system is necessary to maintain the strength of the material at a maximum. A compromise between the two opposing considerations is necessary, based on appreciation of the relative importance of waste material in the runner system, size of the gate scar and the effect on strength. Many small mouldings do not require the full strength possible with a BMC, and in such

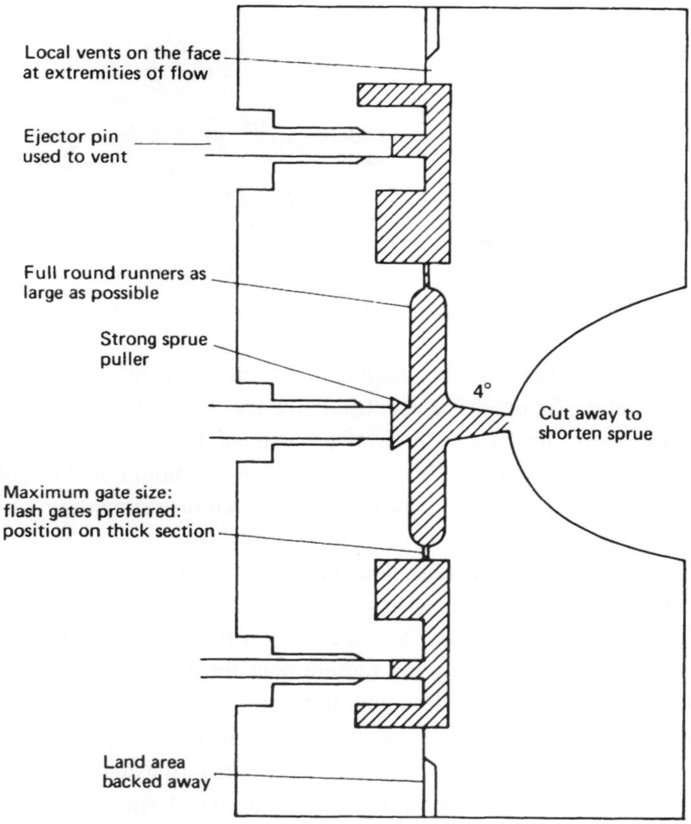

Figure 7.15 General features of a BMC injection mould.

instances small-size runners and small gates can be used. For large mouldings, especially those requiring maximum strength, a direct sprue feed of substantial size will be the best choice. Where maximum strength and reasonable economy will both be required, then those factors having the greatest significance should be given the most consideration.

It is important to bear in mind in this connection that full, round runners are preferable, because they provide the least resistance to flow and cause the least fibre damage.

Sprue lengths should be kept to a minimum, not only to give the best strengths, but also to ease extraction of the sprue, which might otherwise prove troublesome with low-shrinkage materials. A minimum taper of 3° included is required. This is particularly important when moulding small, thin-section parts that can be expected to cure in a very short time, as, in such instances, a very much heavier sprue size would dictate a longer cure and reduce the economic benefits of injection moulding.

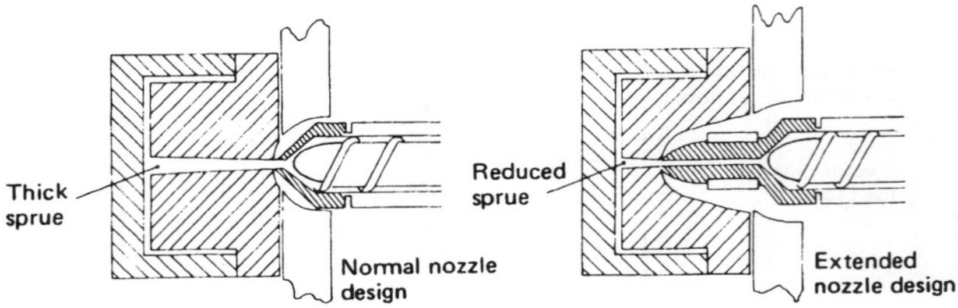

Figure 7.16 Use of an extended nozzle to reduce sprue size and cure time.

Short sprues may often only be obtainable by cutting away the back of the mould to allow entry for extended nozzles on the barrel (see Fig. 7.16). If this arrangement is used, then the position of the extended nozzle is such that a considerable amount of heat will be picked up from the mould, and therefore a separate jacket is required on the nozzle to maintain the temperature at the same level as that in the barrel.

Gate design may be any of those conventionally used for other materials, although careful consideration needs to be given to the effect of the gate on fibre degradation and orientation. The exact size of the gate, and particularly its thickness, is often dictated by the requirement for ease of finishing, but general thickness should not be less than 0.05 mm.

The gate position is important, and mostly it will best be placed on the heavier section of the moulding, as this is best filled first and the thinner sections last, thus maintaining pressure on the material during filling of the mould. It is often advantageous to have a back taper on the tab gate to avoid breakage of the gate within the moulding. This can be particularly troublesome with fibre-reinforced materials. A sprue puller should be provided, and this should be tapered, or of a Z design, to permit a strong pull, which is particularly necessary for materials of very low shrinkage.

As with compression moulds, the ejector pins are also used for venting where this is locally required; venting is often needed on the mating face of the two halves of the mould, so as to allow the escape of entrapped air at the extremities of flow (see Fig. 7.17). This type of vent is best provided by cutting a very shallow groove across the landing face of the tool. This needs to be kept very shallow (0.05–0.10 mm), and the length needs to be kept to 1.0–1.5 mm, so as to avoid premature curing of the material in this very thin section.

Where the technique of injection–compression moulding is necessary, for example to avoid weld lines or orientation, the mould should be designed with a cavity exactly as shown for compression moulds (see Fig. 7.18). This is necessary to contain the material that is injected into the mould while the mould is in the partially open position. Then it is often possible to arrange for

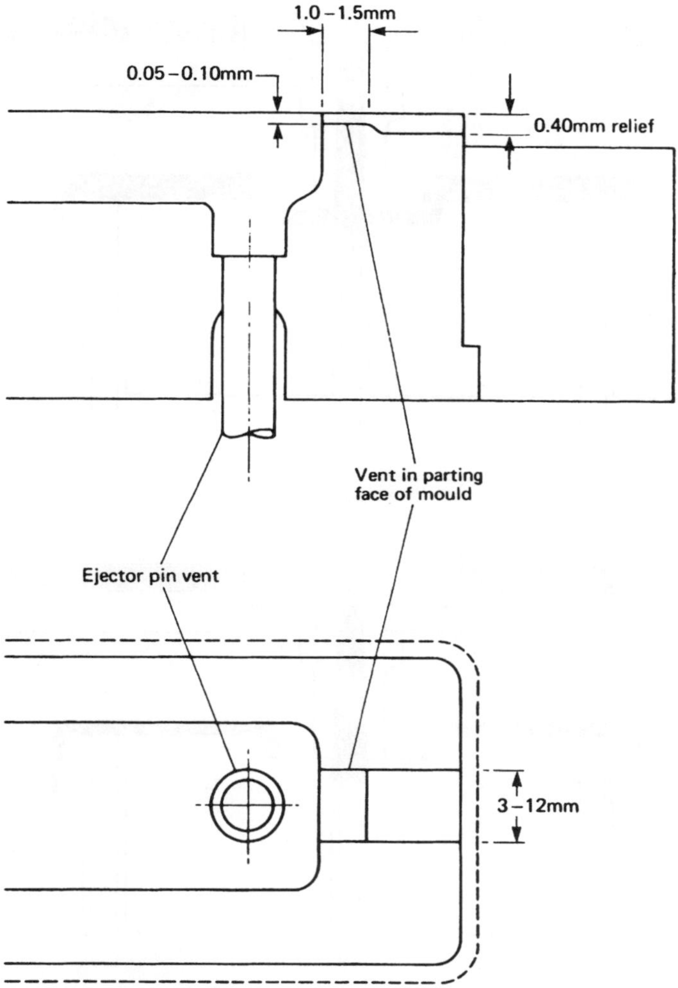

Figure 7.17 Design of venting on the parting face and a boss.

the feed system and gate to be cut off on the compression stroke, thus eliminating the cost of the gate-removal operation.

The use of warm runner moulds in multi-cavity tools for small-section parts, where the weight of the material in the feed system may exceed that of the components, is an economically attractive option. In warm runner moulds, the whole of the feed system is maintained at a lower temperature than the mould (usually 80–100°C) so that this material remains 'live' and capable of being injected in the next shot. The essential features of such a system are shown in Fig. 6.14.

Although the tooling for this technique is relatively expensive, this cost

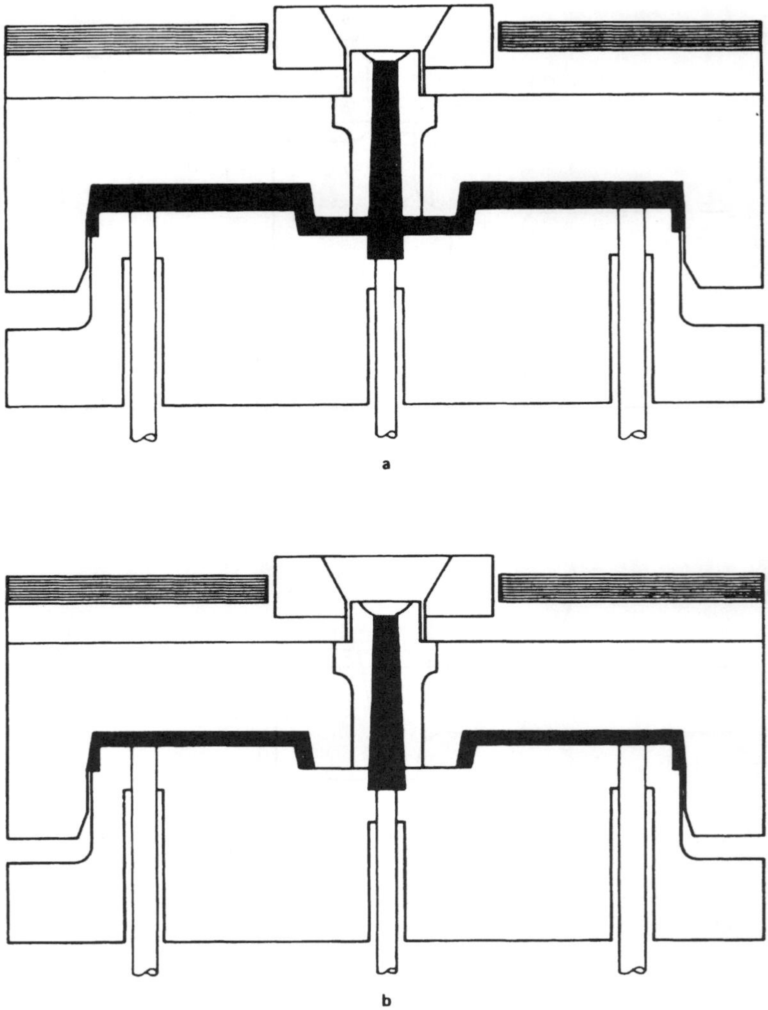

Figure 7.18 Injection–compression moulding of BMC. (a) Mould in open position after injection of BMC. (b) In closed position after compression phase, showing completed moulding and cut-off sprue.

will be offset by the saving of material. Operation of such tools is not particularly difficult with BMCs because of their very easy flow and long life below the critical temperature of the catalyst. This is quite contrary to the warm-runner injection moulding of other thermosettting materials, where the operation is much more critical, and set-up of the materials in the runner system can only be avoided by very careful control of the manifold temperature and the use of specially selected materials.

7.2.6 Injection moulding of large mechanical components (the ZMC process)

During the mid-1980s a significant project was mounted in Europe to improve the mechanical properties of large injection moulded BMC parts with the principal aim of producing the rear hatch door of a new range of Citroën vehicles.[62] The project involved a collaboration of materials supplier, moulder and end-user, and became known as the 'ZX Project' after the principal vehicle model. The moulding process became known as the 'ZMC process'.

Because of the size of components to be moulded, the inherent problem of weld lines around the window and the vital need to optimize mechanical properties, the injection process arrived at was to have split line injection from a single fan gate. However the flow length requirement of up to 2 metres placed severe demands on the rheological behaviour of the moulding compound.

The development and optimization of the BMC formulations was arrived at by a series of statistically designed experiments, with the most critical elements being the choice of catalyst system to avoid pre-cure during mould filling but have an acceptable overall cycle time, and the choice of glass fibre to have optimum level of filamentation to give a good balance of flow characteristics, surface finish and final part strength. Table 7.8 summarizes the relative significance of catalyst choice, fibre characteristics and compound mixer blade speed.

The developed injection moulding machine was an important part of the system. The chosen approach was to use a screw system to transport, homogenize and pre-measure the injection charge. The injection is then made by a plunger type displacement of the complete screw group which is resident within a main injection barrel (see Fig. 7.19).

From Fig. 7.19 it can be seen that this machine construction attempts to combine the advantages of both a screw and a plunger system. The proven

Table 7.8 Synthesis of results of study of compound optimization for ZX hatchback situation

	Parameters			
Factors	Catalyst type	Fibre bundle tex	Fibre length	Mixer blade rotation speed
Flow	++	—	—	—
Mechanical properties	—	+	+	—
Surface finish	—	+	—	—

++ very significant
+ significant
— low significance

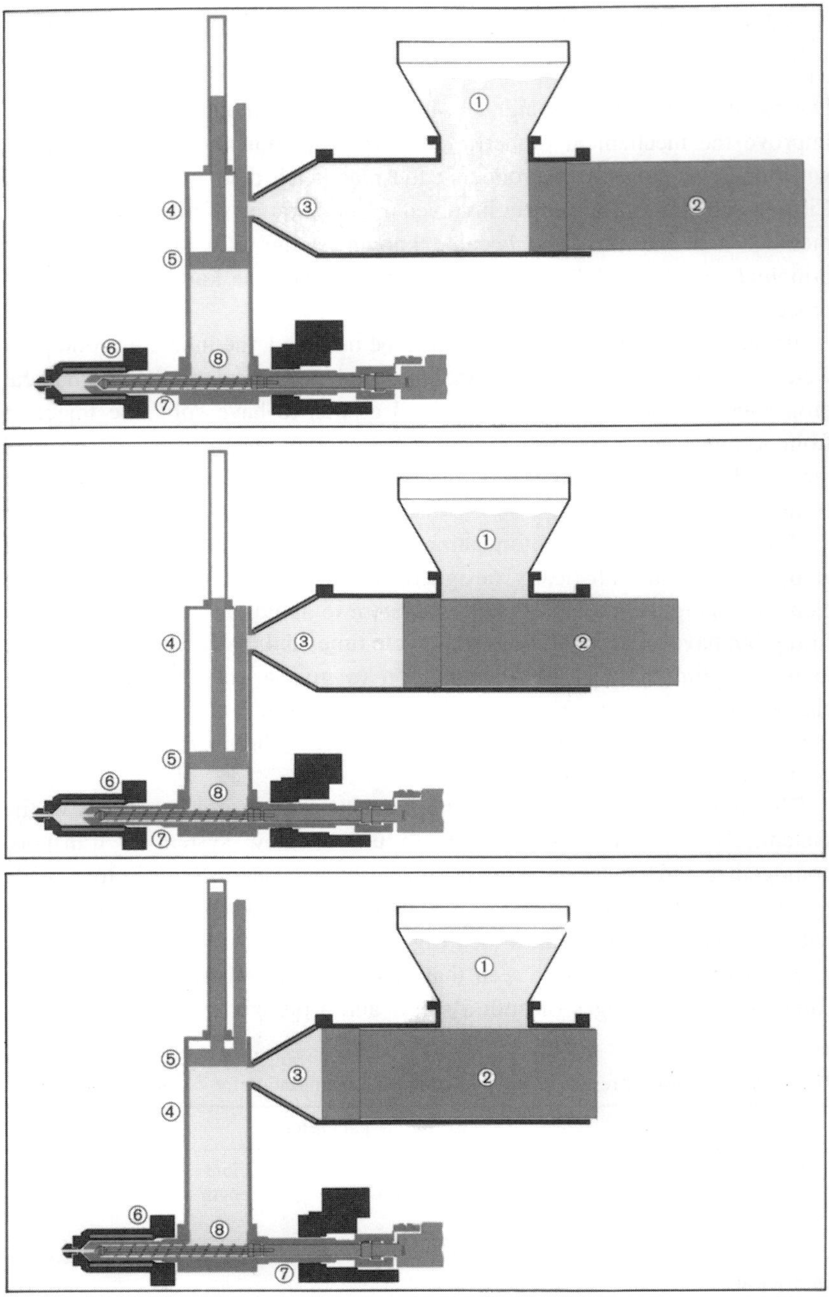

Figure 7.19 Injection unit for the ZMC process. (Courtesy of Billion SA).

Table 7.9 An example of a typical formulation for the ZMC process

Paste		Compound	
Unsaturated polyester resin	55–65	Paste	80
Low profile additive	35–45	Glass fibres (25 mm)	8
Styrene	0–5	(12 mm)	12
Catalyst	1.2–1.7		
Zinc stearate	3.5–4.5		
Calcium carbonate	200–250		

efficiency of screw transfer of the material in terms of achieving good homogeneity and precise measurement of the injection charge is exploited. Then a plunger movement is chosen for injection of the charge into the mould, since this system can have advantages in reducing glass fibre degradation to lower levels than may be possible with direct screw injection, where the necessity of a check valve can influence fibre abrasion and damage.

The compounds evolved for the process have relatively long glass fibres, in the length range from 12 to 25 mm and with compound viscosity adapted to the injection machine characteristics. An example of a typical formulation is given in Table 7.9. The mechanical properties obtained can, in general terms, be said to fall between BMC and SMC. With this potential the process shows itself as worthy of further exploitation.

7.2.7 *Thick moulding compounds and continuous impregnated compounds*

A more recent development in the formulation and manufacture of polyester compounds has been the emergence of two new families known as thick moulding compounds (TMCs) and continuous impregnated compounds (CICs).

Both new materials are manufactured by a continuous mixing process which enables high filler and glass loadings to be achieved with fast machine throughput (Figs 7.20 and 7.21).

The compounding phase (and equipment) is more or less identical for both the TMC and CIC materials. The pre-mixed resin paste is distributed onto two large counter-rotating mixing rolls. Glass fibres (chopped for TMC and cut rovings for CIC) are then distributed onto the rolls and efficient mixing and impregnation takes place as the resin–glass compound passes through the restriction between the rolls.

The two processes then differ in the subsequent stages. TMC compound is sandwiched between polyethylene film and passed through impregnation rolls to give material in a format very similar to traditional SMC. CIC compound is removed from the mixing rolls by doctor blades and transported by a screw or plunger mechanism into boxes or on drums to give a material with similar physical characteristics to traditional BMC.

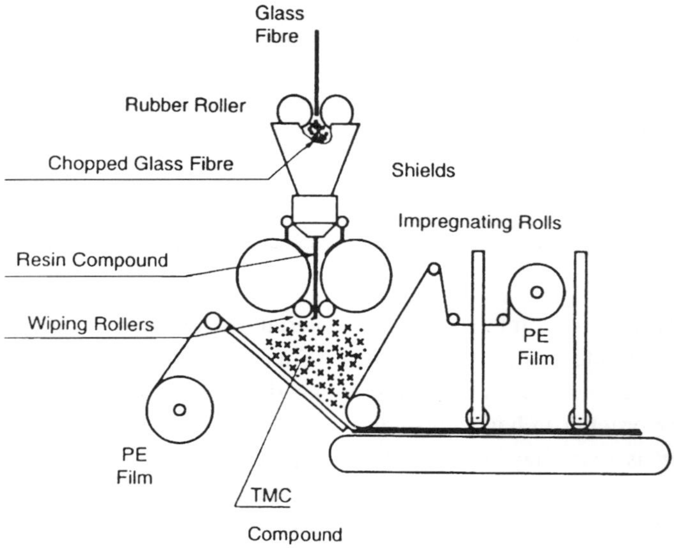

Figure 7.20 TMC compounding process. Reproduced by kind permission of Carl Hanser Verlag from **Kia** (1994), *Sheet Moulding Compounds.*

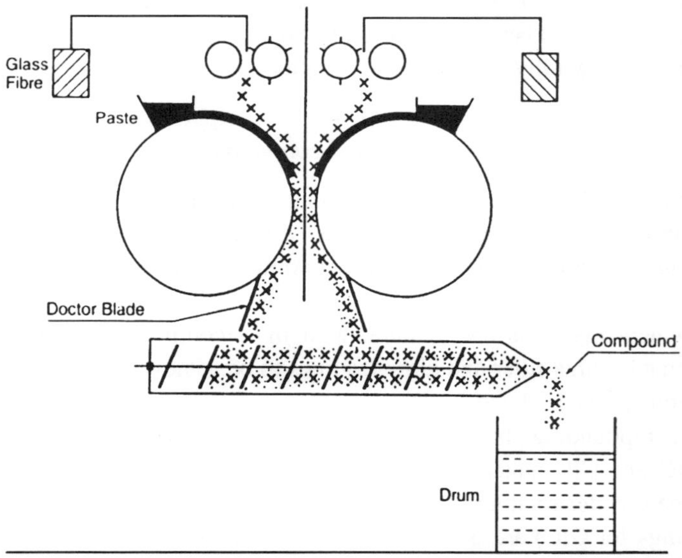

Figure 7.21 CIC compounding process. Reproduced by kind permission of Carl Hanser Verlag from **Kia** (1994), *Sheet Moulding Compounds.*

Because of the simplicity of these processes, it is possible to obtain good wetting of glass fibres while avoiding glass fibre damage. It is also possible to consider compounds with higher glass fibre contents up to and beyond 25%. A comparison of some of the main physical properties of TMCs, BMCs and SMCs is given in Table 7.10.

Table 7.10 Comparison of properties for some representative BMC, CIC, TMC and SMC compounds

Compound	Glass content (%)	Flexural strength (MPa)	Flexural modulus (GPa)	Tensile strength (MPa)	Charpy impact strength (kJ/m²)
BMC	20	100	11	42	22
CIC	20	120	11	58	41
TMC	25	160	10	80	82
SMC	25	170	10	70	90

7.3 Sheet moulding compounds (SMCs)

7.3.1 Basic description and manufacture

SMCs are qualitatively similar to bulk moulding compounds. They are based on unsaturated polyester resins with a styrene monomer, and are filled with the same types of glass fibre, fillers, catalysts, pigments, etc. They differ from BMCs in the proportions of these ingredients, the process by which they are manufactured and the form in which the material comes. The sheet format of these materials means that they are tailored to be processed principally by compression moulding. In the opinion of the author, SMCs do not really lend themselves to injection moulding.

Whereas a BMC is compounded very simply in a Z-bladed mixer, an SMC is formed in a machine (see Fig. 7.22) consisting of a series of rolls that are devised to bring the ingredients together and consolidate them into a sheet sandwiched between two layers of polyethylene film. The two sheets of polyethylene film are brought into the machine at different levels and are coated with a layer of polyester resin that has previously been filled with the particulate mineral filler together with the catalyst inhibitor and a thickening agent. Glass fibre is chopped and deposited on one of these coated films and then sandwiched by the other. In this way the glass becomes consolidated with the filled resin into a homogeneous sheet, which is then rolled up at the end of the machine. If this process is compared with the process for the manufacture of TMC described in the previous section, and illustrated in Fig. 7.20, it can be seen that the TMC process makes use of an additional pair of mixing rolls to assist the continuous paste/glass fibre impregnation.

Thickening agents are an important ingredient in an SMC formulation, in that they cause a thickening process to occur over a predetermined period. The most common agents used are the oxides or hydroxides of magnesium or calcium, alone or in blends, with concentrations between 0.5 and 3.0%. These cause the polyester resin to partially gel and thus transform the sheet into an essentially tack-free, handleable material. This thickening process takes several days to reach completion at a maturation temperature which is usually about 25–40°C. A typical thickening curve for an SMC moulding compound

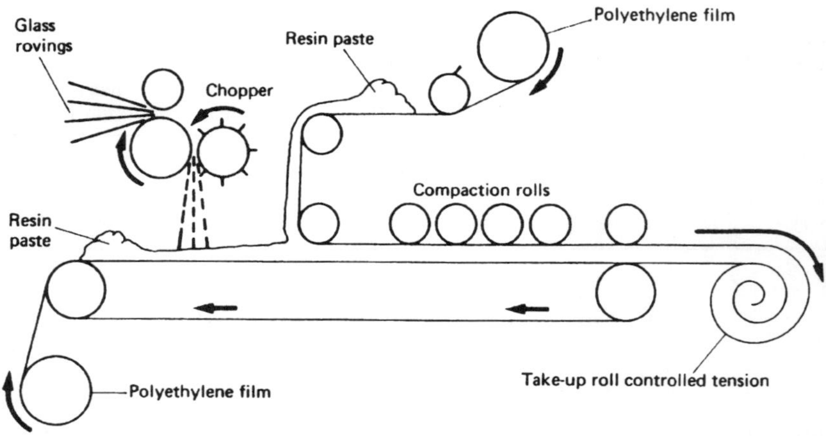

Figure 7.22 SMC compounding process.

is given in Fig. 7.23, which shows the viscosity increase against time and also indicates the period in which the material can be considered as having reached a satisfactory condition for moulding. The curve presented is not intended to be absolute, but indicates the general thickening response obtainable. To obtain a thickening curve for a specific formulation, laboratory trials must be carried out.

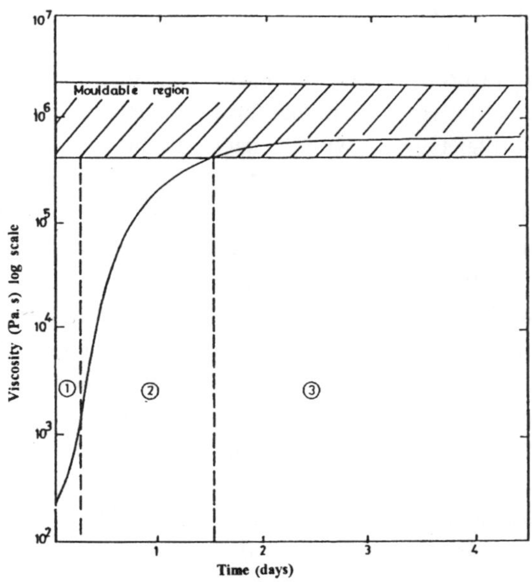

Figure 7.23 Typical thickening curve for an SMC showing the mouldable region.

The dispersion of thickening agents in the compound is a critical point, to ensure that the thickening effect is fully homogenized. It is often considered beneficial to pre-disperse the thickening agent in liquid form to assist incorporation into the SMC paste. Even with this refinement, it is not unknown for variation in thickening rate to occur in routine production, since variations in other compound elements such as resin, fillers and moisture levels can all significantly influence this process.

Apart from any convenience in the form of the material, which may be more suitable for large thin mouldings, the most distinctive feature of an SMC compared with a BMC is its significantly higher strength. This arises from the nature of the process, because, whereas the mixing process for BMC manufacture causes entanglement and some damage to the glass fibres, to the detriment of the mechanical strength, the SMC manufacturing process causes little or no fibre degradation or entanglement. Further, incorporation of larger amounts of glass fibre is possible with SMCs.

SMCs mould less well than BMCs on complicated shapes, but are particularly suited to large shell-type mouldings (of which automotive body parts are a particularly good example) where the very much higher strength provided by an SMC is essential.[63]

The application of SMCs is very wide and covers many uses where high strength is a prime requirement. It has found considerable use in car body parts, truck cabs, fascias, as well as many electrical and large structural components where BMCs are not considered sufficiently strong (see Figs 7.24 and 7.25).

Most of the properties of SMCs are similar to those provided by BMCs, except for the very much higher strength (see Table 7.11). The strength obtained varies with the glass content, which may be as high as 50%, and it is even possible to produce grades with unidirectional reinforcement to give the highest possible levels of strength. However, moulding characteristics become poorer at the higher glass contents, because of the restriction to flow provided by this change.

Satisfactory moulding characteristics are not fully developed until the SMC has matured adequately, that is, the thickening agent has reacted with the polyester resin. However, this process, although substantially complete after a few days, does continue very slowly beyond that time and the sheet will eventually become too hard to mould satisfactorily. This is often the limitation of the storage life, even though the material may still be capable of being moulded under much higher pressures and capable of reaching a satisfactory cure. It is particularly important, therefore, to store the material at cool temperatures and check from time to time that it is still of adequate moulding quality.

For readers who wish to follow this class of materials in greater detail than provided in this chapter, other works are recommended.[64,65]

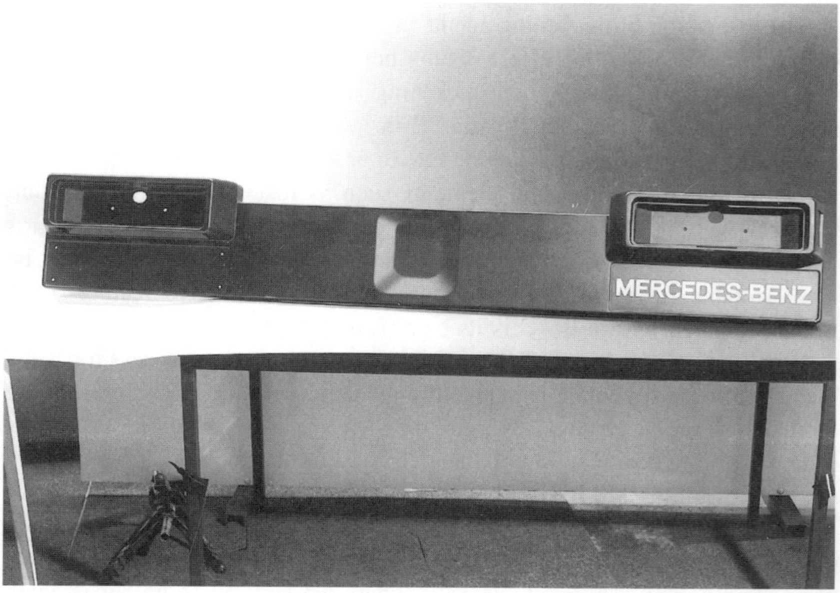

Figure 7.24 An example of an SMC automotive application. (Courtesy of Maschinenfabrik J Dieffenbacher GmbH & Co.)

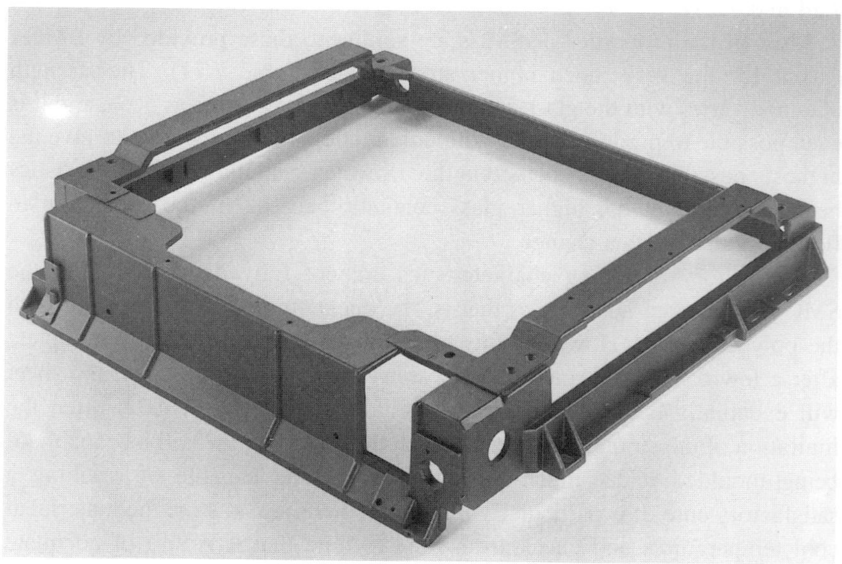

Figure 7.25 An example of an SMC general application. (Courtesy of Maschinenfabrik J Dieffenbacher GmbH & Co.)

Table 7.11 Properties of SMC compounds with different glass fibre contents

Property	Units	Glass fibre content (%)		
		15.00	20.00	25.00
Density	g/cm^3	1.75	1.85	1.95
Mould shrinkage	%	0–0.2	0–0.2	0–0.2
Tensile strength	MPa	40–52	45–65	60–80
Flex. strength	MPa	80–110	95–130	120–155
Flex. modulus	GPa	8–9.5	9–10.5	10.5–12
Notched impact	kJ/m^2	30–50	38–60	55–70
Surface resistivity	Ωcm	10^{12}	10^{12}	10^{12}
Deflection temperature (°C)	1.8 N/mm^2	> 200	> 200	> 200
Coeff. of therm. exp	10^{-5} K^{-1}	2–3	2–3	2–3
Water absorption	mg	20–40	20–30	20–40
Fire retardency	Category	None	None	None
Oxygen index		21–50	21–50	21–50

7.3.2 *Presses*

The basic information on press requirements given for BMCs earlier in the chapter applies equally to SMCs as far as pressure and controls are concerned. However, many parts made from SMCs, especially those for transport applications, are required to have a particularly good surface finish, free from porosity and capable of being painted without pretreatment of the surface. Where this is a requirement, it has been found necessary to use fast-closing presses to eliminate pre-cure and ensure good consolidation of the material, so as to avoid surface porosity. Adequately high moulding pressures are also an essential requirement in such applications. The specification for such a press is as follows.

Closing force:	capable of 100 bar pressure on largest mould area.
Fast closing speed:	600 mm.s^{-1} at rated closing force.
Slow closing speed:	continuously variable from 25 to 0.25 mm.s^{-1}.
Control:	accurate automatic alignment of platens to ±1 mm parallelism.

The very fast closing speeds specified are no more than a good modern small press would provide, but larger presses, such as are often needed for large body components, do not usually match this specification, and it has therefore become necessary to emphasize this particular need. Equally there is a tendency, because of cost, to skimp the total moulding pressure made available on large presses, but this can be disastrous to the surface quality of mouldings made from SMC. Although fast closing speeds are emphasized, it is still necessary to have provision for a slow close towards the end of the stroke, and the position at which this change in the closure occurs and the rate of close both need to be controllable.

On very large presses it is often the practice to provide hydraulic sensing devices on the four corners of the platen to ensure that when the platens come together they are quite parallel. Deflection of the platen on large areas is quite common and is aggravated by the laminar stack of SMC sheet which forms the charge to the mould. These laminations can slip sideways and resist the closing of the tool asymmetrically. A press modified in the way described will react to this and keep the platens parallel, thus ensuring a more consistently accurate part.

7.3.3 Moulding conditions

Because of the close similarity between SMCs and BMCs, the moulding technology is similar.

The range of moulding temperatures and pressures used for SMCs is the same as that previously defined for BMCs but rather lower temperatures (typically 135°C) and higher pressures (typically 70 bar) are generally preferred. Under similar conditions, more pressure than for BMCs is necessary to consolidate SMCs and make them flow. Lower temperatures are also desirable to extend the flow life. Removal of air entrapped between the layers of SMCs may be difficult, and is favoured by the longer gelation times at the lower moulding temperatures.

The material cures rapidly, but the cure will be affected by various pigments, and is especially slowed by carbon black.

It is always essential to close the mould rapidly to the point where the top face of the mould comes near to the charge of material, and, from this point onwards, the rate of closure should be slowed to allow gentle consolidation of the layers of material and easy escape of air entrapped. The precise rate of closing has to be determined by trial, and should be the maximum speed consistent with obtaining a moulding free from blisters and excessive flow lines. Procedures such as dwelling or breathing are not suitable for use with SMCs because they spoil the surface finish and introduce surface porosity.

The laminar character of SMCs makes charging of the mould a particularly important consideration. The best practice is to cut pieces of the sheet to a size roughly equal to 70% of the area of the mould. There are two considerations in this procedure. The first is that if the material does not have to flow much to reach the extremities of the mould, then the geometry and arrangement of the glass fibre is not disturbed and the maximum strength will be obtained. The second consideration is that, unless the material is made to flow some distance, it will not consolidate well and the surface appearance will be inferior. It is for this reason that most modest-size mouldings are charged with pieces of material equal to about 70% of the moulding area. However, very much larger parts may use a higher proportion than this, whereas small or complex parts may necessarily have to be charged with material of much smaller size. This is also true of deep draw mouldings.

These factors need to be carefully taken into account, because very

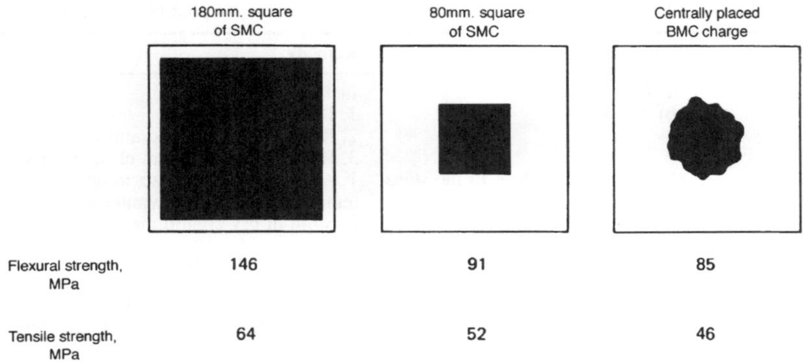

Figure 7.26 Effect of mould charge size and distribution on the strength of SMC and BMC mouldings (both SMC and BMC contain 20% glass fibre).

considerable strength reduction will occur if the flow path is long and the glass fibre much disturbed. Figure 7.26 gives results of an experiment that illustrates the effect of charge size on the strength of an SMC moulding. A 200 mm square by 3 mm thick panel moulding produced from pieces of SMC measuring 100 mm square gave mechanical properties very substantially less than those obtained when the moulding was made from pieces of SMC cut 180 mm square. Manufacturers' data nearly always correctly quote the mechanical properties of SMCs on specimens prepared in the latter way, whereas many mouldings made commercially will either accidentally or by force of circumstances be made from small pieces of SMC, and these will have much lower properties, more nearly equal to the performance of a BMC part of similar glass content.

It has also also been found that, even where most of the charge of SMC is of the optimum size, introduction of small off-cuts within that material can introduce very serious reductions of strength, owing to the laminar nature of the flow. Taking all these factors into account, it is important, therefore, to pay particular attention to using the optimum size of SMC charge, or, if this is impossible, to make allowances for the inevitable lower strength.

The common problems in moulding SMCs are listed in Table 7.12, together with their causes and possible remedies.

7.3.4 Mould design

The designs of compression moulding tooling described previously for BMCs are equally valid for SMCs. Vertical flash moulds are even more to be recommended for SMCs, as the provision of positive pressure on the moulding is essential to good quality, particularly with respect to surface finish and freedom from porosity. For this reason, there is a tendency to make

Table 7.12 Guide to SMC moulding faults, with causes and possible remedies

Fault	Cause	Remedy
Blisters	(a) Under-cure	Increase cure time
	(b) Air trapping	1. Reduce charge size 2. Reduce mould temperature 3. Reduce speed of final close of mould
	(c) Exposure of sheet to moisture	1. Allow packages to warm up to shop temperature before unwrapping 2. Store in dry conditions
Poor gloss	(a) Under-cure	1. Increase cure time 2. Increase mould temperature
	(b) Lack of pressure in cavity	1. Increase moulding pressure 2. Ensure mould is reasonably positive and flash clearance is even 3. Increase speed of closure
	(c) Poor alignment of platens with large moulds	Use suitable press
Soft mouldings	(a) Under-cure	Increase cure time
	(b) Material exceeded its storage life	Use fresh material
Local burning	Dieseling of compressed gases	1. Reduce speed of closure 2. Provide adequate venting
Internal voids	(a) Sheets not consolidating	1. Ensure removal of polyethylene film from sheet 2. Increase cure
	(b) Material too hard — exceeded storage life	Use fresh material
	(c) Air trapping	1. Reduce speed of closure 2. Reduce charge size 3. Reduce mould temperature
Surface porosity	Air trapping due to: (a) Pre-cure	1. Increase speed of closure 2. Reduce mould temperature 3. Eliminate dwells and breathers
	(b) Excessively hard material	Use fresh material
	(c) Poor flow	1. Adjust mould temperature 2. Use different grade
	(d) Unsuitable mould design	1. Introduce local venting 2. Use correct flash line design
	(e) Inadequate pressure	Increase moulding pressure
Cracking on ejection, or poor mould release	(a) Uneven operation of ejection	Modify mould
	(b) Undercuts or poor finish on mould	Improve mould finishing
	(c) Uneven arrangement of ejector pins	Modify mould
	(d) Under-cure	Increase cure time
	(e) Mould temperature excessive	Reduce temperature
Swollen white patches	Material exceeded its storage life	Use fresh material
Poor strength	(a) Loading of material to mould	1. Increase size of sheet charged to the mould 2. Eliminate small off-cuts
	(b) Geometry of mould creating unfavourable flow pattern	Unavoidable
	(c) Weld lines next to cored holes	1. Relocate charge of material 2. Machine holes subsequently

SMC moulds even more positive than is usually the case for BMC, and this is most often effected by increasing the depth of the parallel fit on the vertical flash cavity. The length of this parallel fit can be up to 12.5 mm, but it is still desirable to maintain a 2° taper beyond that to allow easy escape of flash.

A further feature that is sometimes practised to achieve particularly good consolidation is to use the mould without pressure pads so that the full pressure of the press is applied to the moulding. Although this undoubtedly has some benefits and improves surface finish and gloss particularly, the mould may be put at risk, and alternative safety measures should be adopted.

7.4 In-mould coating process (IMC)

With the widespread use of SMCs for vehicle exterior panel applications, a complementary process has been developed to allow the paint priming of the panels in an integrated process operation which takes place in the mould. This process, and the associated priming lacquers, have been developed by the Sherwin Williams Company in America, and successfully exploited not only for SMC parts but also for some BMC parts.[66]

The most efficient version of the process involves the high pressure injection (at 200–400 bar) of a charge of thermosetting paint or lacquer which is made to cover the surface of the formed SMC component (see Fig. 7.27). The timing of the lacquer injection is during the phase of curing of the SMC component, with the injection pressure being sufficient to allow the lacquer to flow over the surface of the component, but still avoid opening of the mould to an amount that would cause flashing problems. This is possible because, at the time of resin injection, the SMC part is not fully cured and is therefore still compressible.

Because the pressurized injection of the coating lacquer causes the mould to open by overcoming the clamping force it is sometimes referred to as the injection opened coating (IOC) process. An alternative approach, which operates at lower lacquer injection pressures, is to remove the mould clamping force and open the mould slightly when the component has reached the appropriate stage of cure, then inject the coating lacquer and re-apply the full clamping force again. This method is referred to as the press opened coating (POC) process.

The process not only eliminates a separate priming phase, but is also effective in eliminating any residual surface porosity in the surface of the SMC part. The coating lacquers were originally two-component unsaturated polyester and isocyanate formulations, but are more commonly one-component systems bases on free radical cured unsaturated oligomers and monomers with excellent adhesion to the SMC substrate.

Although most extensively used for large SMC panels, the IMC process can also be considered for some BMC parts (injection or compression) where

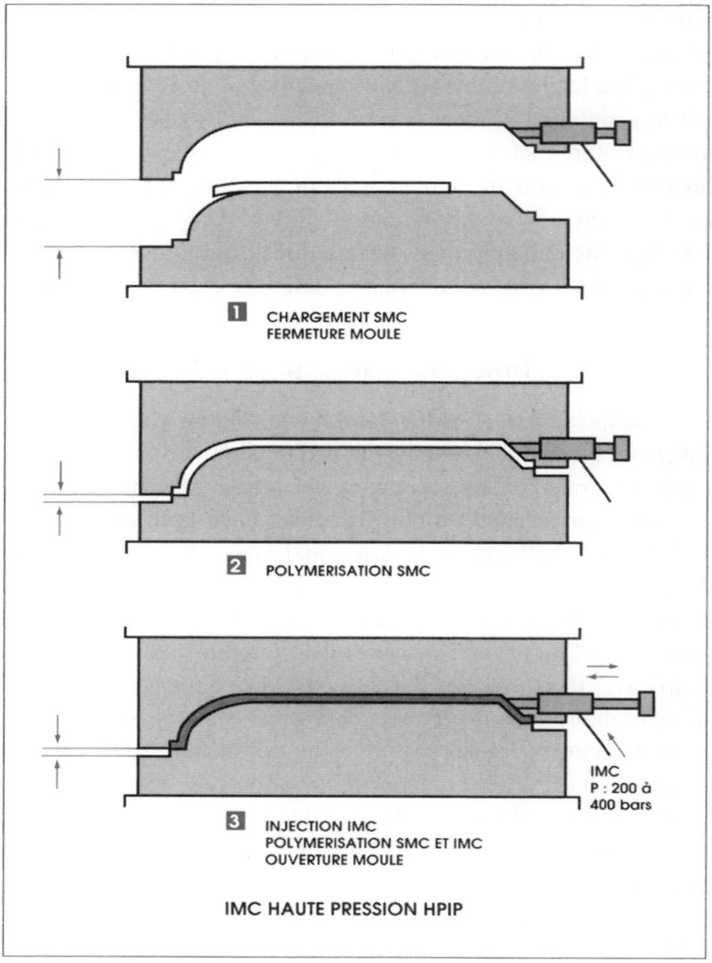

Figure 7.27 High pressure In-Mold Coating process.

the parts are relatively large and simple and have special finishing requirements.

7.5 Reprocessing of BMCs and SMCs

The growing importance of BMCs and SMCs in terms of mass volume applications has been very striking over a considerable period, and has led to a focussing of attention on the issue of waste reprocessing.

In Chapter 1, mention was made of the German Federal Government inspired ERCOM plastics disposal and recycling project, which has demonstrated the potential for tackling this problem. The fact that thermosetting materials cannot be directly recycled, because their moulding process is irreversible, does not necessarily imply that useful and economic solutions to the problem cannot be found.

Based on the studies carried out with reclaimed BMC and SMC components from motor vehicle dismantlers, recyclates have been obtained for use as fillers in thermoplastics materials for the production of non-aesthetic automotive components such as covers, housings and wheel arch liners. Milled SMC recyclate, in particular, can retain sufficient fibre integrity to be used as a reinforcing filler, with particle sizes of 2–4 mm, and then using 10–30% recyclate to replace traditional glass fibre reinforcing materials.

With worldwide volume consumption of BMCs and SMCs now in the region of 1×10^6 tonnes annually, this development is important on both economic and environmental grounds.

CHAPTER 8

Reactive injection processing of polymers

P D COATES AND A F JOHNSON
University of Bradford / IRC in Polymer Science & Technology

8.1 Introduction

Reactive processing involves polymerization or chain extension during a fabrication process.[67–74] It is a necessary step in the production of many materials, for example phenolic or unsaturated polyester resins which require heat curing, and elastomers requiring vulcanization. Around 25% of polymer processing is estimated to involve reaction.

This chapter covers reactive injection processing which involves the production of polymeric solids or polymer composites, the final article and the polymer being produced simultaneously in a closed mould from reactants injected into the mould. Reactive injection processes such as resin transfer moulding (RTM) and reaction injection moulding (RIM) and its derivatives differ from conventional thermoset injection moulding in that they require injection of relatively low-viscosity reagents into a mould in which polymerization or chain extension reactions take place; in the case of RIM processess rapid intensive mixing[75–80] of low molecular weight reagents is required, immediately prior to them being introduced into the mould.[70,81,82]

Figure 8.1 shows schematically the family of these relatively low-viscosity reactive injection processes.[67,68,71] The main focus of this chapter will be the RIM process and its derivatives, RRIM and SRIM, RTM processing being referred to where appropriate. The forerunner to the current high pressure RIM technology was a low-pressure process developed specifically for polyurethane (PU) chemistry, largely for low modulus products such as low-density foams and shoe-soling materials, which employed static or dynamic mechanical mixers.[67,68,72] In the high-pressure process, two or more reactants are mixed intensively by high-pressure jet impingement at a precise ratio. The relatively low-viscosity mixture then passes at low pressures into a closed, generally large surface area mould, where polymerization is completed.

The attractions of low process energy requirements[67,71] and relatively low capital costs have ensured an expansion of interest in this process to other polymeric materials and also to the incorporation of particulate or continuous fibre reinforcement for the enhancement of mechanical properties.

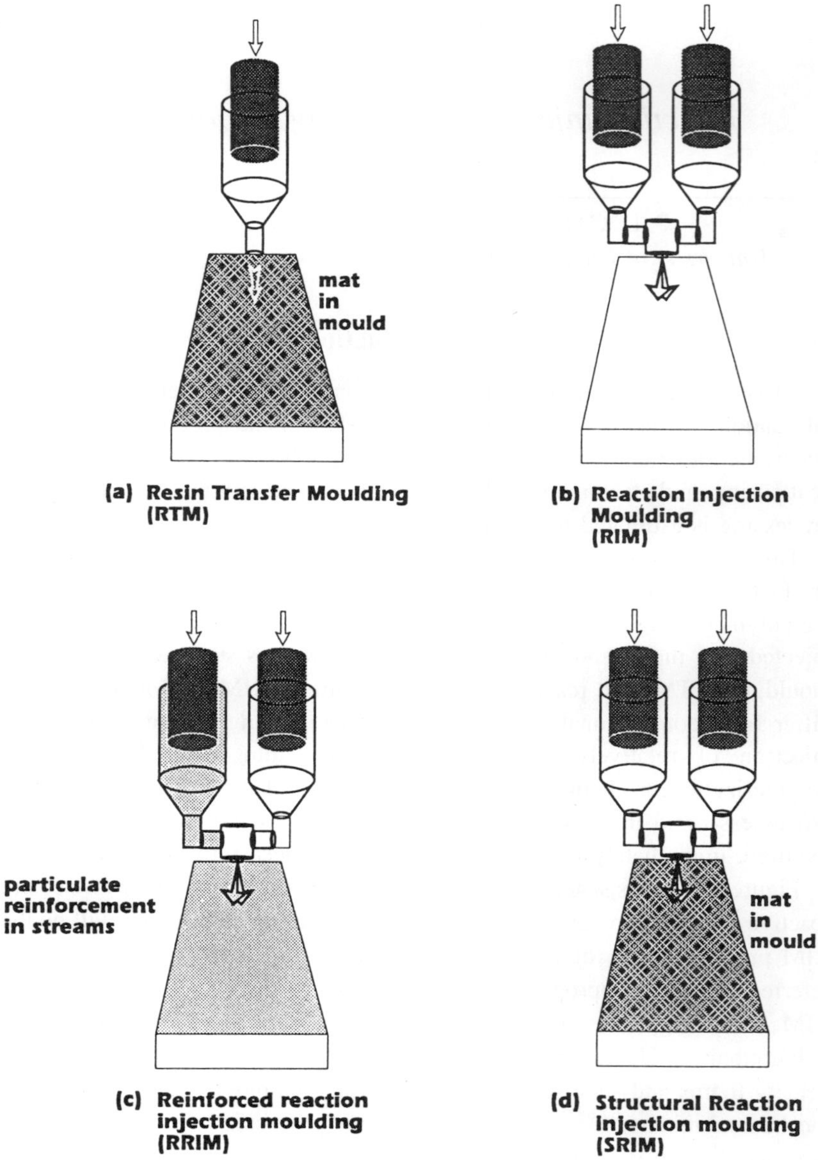

Figure 8.1 Reactive injection techniques.

8.1.1 Resin transfer moulding

RTM (Fig. 8.1a) is the simplest reactive injection process — this provides its commercial appeal and relative success. It has largely been concerned with polyesters injected into moulds containing continuous fibre mats in various physical forms (such as random, woven, or stitch bonded). Prior mixing of the reagents is possible, since the half-life of reaction is very long and therefore no mixing stage is required during the injection process. RTM is almost entirely used for polyester–glass composite products, but both acrylic and epoxy RTM formulations have been explored in our laboratories and elsewhere, and there is current interest in these systems. However, single stream reactive injection processes suffer from several drawbacks, not least being long cycle times, because the criterion for long shelf-life of the pre-mixed reagents militates against rapid in-process reaction. In addition, it is not possible to 'tailor' the process through control of reagent stream ratio, or incorporation of additional streams. RTM process technology tends to be unsophisticated, particularly the dispensing equipment which is often pneumatically actuated.

8.1.2 Reaction injection moulding

RIM technology (Fig. 8.1b) has been in use for many years as a means of handling reagents which react so rapidly that they have to be stored separately prior to mixing, for example the isocyanates and alcohols which when mixed react to form polyurethanes (PUs). Indeed, as noted earlier, RIM technology is actually based upon the requirements for processing PUs, which are still by far the most common RIM products. Other polymerizations have been tailored to fit with the technology developed for PUs, or adaptations of it.[68,73]

In general, two-stream RIM machines are used and the reagents have to be formulated accordingly. Sometimes a third minor stream is used commercially in order to introduce foaming or colouring agents into the reaction mixture (or, as in the case of our laboratories, third and fourth reactant streams may be injected to form polymer alloys). PU formulations are normally highly reactive, and generally require only ambient or near-ambient temperature dispensing, with moulds controlled around 65°C — these relatively low processing temperatures also contributed to the success of PU RIM.

8.1.3 Reinforced reaction injection moulding (RRIM)

RRIM (Fig. 8.1c) was originally developed as a method for the production of polyurethanes with enhanced physical properties through the incorporation of particulate reinforcing fillers in one or more of the reagent streams. In order to process particulate materials incorporated into the reagents, the RRIM processing machinery has to be fitted with pumps which are capable of handling viscous and abrasive slurries. The universal tendency has been to use lance positive displacement pumps. The engineering properties of the products can be greatly enhanced by structural reaction injection moulding

(SRIM) (Fig. 8.1d), that is RIM of a polymer matrix over a preplaced continuous fibre reinforcing mat, as in RTM.

8.1.4 Range of materials for reactive injection processing

Key variables in RIM-type processes are shown in Fig. 8.2: these may be conveniently categorized as those related to materials and those which relate to process technology. Reactive injection processing (in common with other reactive processing operations) involves a complex inter-relationship between these two sets of variables. Controlled reactive processing therefore demands that processing technology and the polymerization chemistry or chain extension reactions are well matched in order to optimize product properties and quality whilst maintaining acceptable production rates.

General comments on materials in this section will be followed by more detailed discussion of selected polymerizations (cf. 'chemical variables' in Fig. 8.2), prior to discussing process flows and process technology (cf. 'machine variables' in Fig. 8.2).

During the late 1970s and early 1980s there was a very rapid growth in interest in the use of reactive injection processing, particularly RTM and RRIM methods, for the production of large surface area, thin products for automobile body panels and bumpers, mainly involving PUs. RRIM has been extended to mould a large variety of polyurethane parts, of increasingly large mass, with a wide range of physical properties, for both exterior and interior automobile use. Particulate or fibrous fillers are usually incorporated to increase the material stiffness and reduce the linear coefficient of thermal expansion of PUs. Brittle reinforcing fillers tend to reduce the product impact resistance and a compromise has to be reached as far as physical properties are concerned (although the use of more ductile high stiffness polymeric reinforcements is being explored).

Because of the advantages and successes of RIM as a polymer processing technique for PUs, RIM techniques are being applied to many other materials. Most of these have relied upon, arguably, relatively minor (but not necessarily cheap) changes in existing processing machinery, with the limiting factor for development being advances in chemistry to provide monomers and catalysts with the correct combination of viscosity and reactivity, and which form polymers with good physical properties and no undesirable by-products.

RIM polymerizations currently include, in order of commercial useage, polyurethanes, polyureas (PUR), polyamides and polydicyclopentadienes. Developments are occuring in epoxy, acrylic and polyester systems, and in polymer alloys. Although the original drive for development of new RIM polymers came from the automotive sector, the materials developed have found other applications. Specific RIM polymerizations are discussed briefly

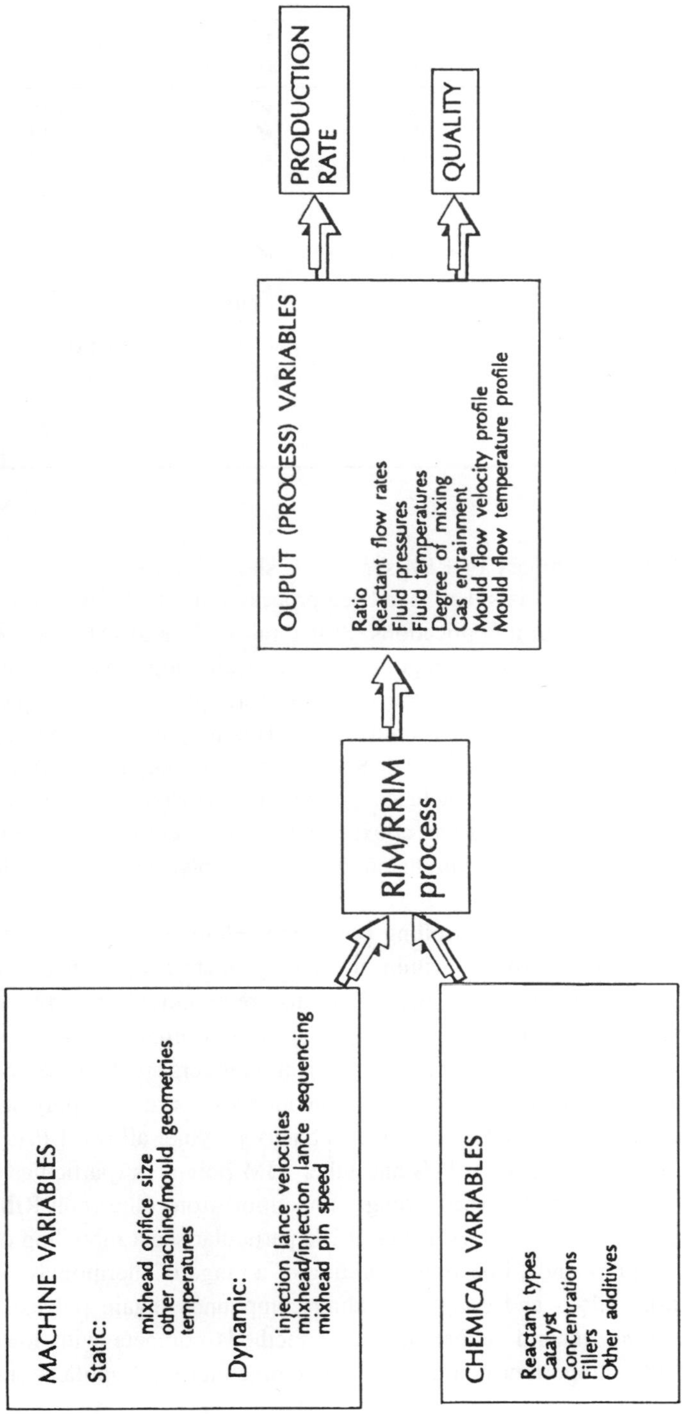

Figure 8.2 RIM processing variables.

Table 8.1 RIM polymers

System	*T* reactants (°C)	*T* mould (°C)	Fully commercial	Useage
Polyurethane	~30–40	~60–90	✔	v. extensive
Polyurea	~30–40	~60	✔	v. extensive
Polyamide ['Nyrim']	~90	~140	✔	
Dicyclopentadiene ['Metton', 'Telene']	~30–40	~140	✔	some
Acrylamates ['Arimax']	~30–40	~120	✔	
Epoxy ['Epikote']	60+	~140	partly	
Acrylics ['Modar']	40	120		RTM
Polyester	60+	~120		
Binary polymer alloys	depends on systems			

in section 8.2 and a summary of RIM systems is shown in Table 8.1.

Recently, the new materials that have been processed by RIM have tended to be polymers for structural applications. Structural RIM or SRIM products represent the current direction of progress in RRIM technology. An essential contribution to the production of these engineering plastics is the pre-placement of reinforcement in the mould cavity. This may be glass matting or, for very high stiffness, carbon fibre or Kevlar. The base polymers used for these advanced applications include polyamides (nylon), polyesters, polydicyclopentadienes, acrylics and epoxies, although recent developments have also been made in rigid polyurethanes and hybrid polyesters which hold promise in SRIM.

An alternative method of controlling polymer modulus is to incorporate rigid organic polymeric fillers within a rubbery matrix by concurrent polymerization of both networks. These materials are commonly referred to as 'interpenetrating polymer networks' (IPNs), which implies that they possess only a single phase and that the component polymers are dispersed on a molecular level. In practice, they may contain two (or more) polymer phases and may therefore best be described as binary polymer alloys (BPAs).

There is competition between PUs and other RIM polymers (particularly for automotive use), and also very strong competition from other non-RIM polymer-processing techniques and materials, in particular sheet moulding of polyester–glass compounds, injection moulding of a range of thermoplastics and thermoplastic alloys and dough moulding compounds (again polyester based). In addition, general rubber moulding methods compete with low-modulus RIM PUs. Important non-automotive commercial RIM applications include structural PU foams.

8.2 RIM/RRIM/SRIM materials

8.2.1 Polyurethanes

The widespread application of these materials[69,72] comes from the diverse nature of their precursors, which allows molecular and morphological tailoring, which in turn gives control of the polymer properties and an ability to meet the requirements of many specific end uses. The physical properties of PUs have been studied far more extensively than any other class of reactively processed polymer. Non-foamed polyurethanes are segmented block copolymers of varying degrees of cross-linking, comprising rigid and rubbery phases. It is this two-phase morphology which imparts the basic properties of rigidity coupled with impact resistance.

Polyurethanes are formed by the reaction between multi-functional isocyanates and alcohols to form the urethane or carbamate linkage. Linear polymers are formed when each of the co-reactants is difunctional. The chemical structure of the hard and soft block segments can be readily changed, as can the relative amounts of hard and soft block component in the polymer. When the hard block component is in excess the soft block component improves the impact resistance of an otherwise brittle material. When reactants have three or more functional groups, cross-links are introduced into the polymer structure, which serve to stiffen the matrix. The overall degree of cross-linking is dependent on the proportion of reactants of higher functionality. The alcohols generally comprise a polymeric diol (molecular weight typically 500–3000) and a low molecular weight 'chain extender', such as butanediol or ethylene glycol. Diamine chain extenders can also be used, to produce a polyurethane urea.

Polyurethanes with properties tailored to a very wide range of applications can be produced because of the following factors:

- the ease with which the degree of cross-linking and the ratio of hard to soft block can be manipulated;
- the lack of low molecular weight by-products, which would have detrimental effects on physical properties;
- a vast range of available isocyanates, polyols and fillers;
- control of foaming and gel time through catalysis.

Catalysts can be selectively used to control the gel time and this can be reduced to as little as one second. The two main classes of catalyst used in PU formation are tin(II) compounds and tertiary amines. The choice of catalyst can influence the structure of polyurethanes, since each type of catalyst favours a different set of reactions. Amines favour the reaction of isocyanate with water which is the source of foaming. Tin catalysts favour the polymer-forming reactions between alcohol and isocyanate.

Reaction injection moulded PUs have a dense skin and microcellular core. This is a consequence of carbon dioxide production when traces of moisture in the alcohols react with isocyanate, or the result of dissolved and entrained gases within the reagents being liberated due to the large heat of reaction generating a substantial temperature rise within the interior of the moulded part. Polyurethanes cure from the centre (high temperature region) outwards, generating gas in the regions of fast reaction and compressing material in the region of the mould wall. The 'skin' is advantageous for applications requiring good surface finish and permits automotive 'class A' surface finish to be obtained, whilst the microcellular core reduces weight without severely decreasing strength and stiffness. Careful control of foaming is essential — closed-cell foams are required if good properties are to be maintained. The heat of reaction in PU formation limits the maximum thickness of parts that can be formed: in very thick mouldings, heat generated would not be conducted to the cooled mould wall and subsequent temperature rises would lead to thermal degradation.

Almost all RRIM PU formulations for automobile application are based on derivatives of 4,4'-diphenylmethane diisocyanate (MDI) and polyether or polyester polyols. These are all ultimately derived from crude oil. Alternative materials based on furan diisocyanate and on polyols derived from agricultural waste are products of renewable chemical sources and may gain in usage should crude oil prices become high. For car body panelling applications, polyether-based PUs are used because of the higher hydrolytic stability of polyethers compared with polyesters. These materials are also cross-linked in order to improve the high temperature modulus characteristics. However, it is impossible to achieve an acceptable level of physical properties for engineering applications without the incorporation of reinforcing fillers. The PUs used in automobile panels usually contain approximately 25% hammer milled glass. The reinforcing filler has a significant effect on the linear coefficient of thermal expansion and on the heat sag properties of the materials. Typical cycle times in batch reactive processing of PUs for automotive use are of the order of 1–2 min.

Polyurethanes find application in many areas other than the automobile industry: for example business machine housings, sports goods, furniture, insulation materials, agricultural equipment and footwear.

Recently, the possibility of recycling of PUs by glycolysis has been demonstrated — in some cases this can lead to recovery of a polyol which is sufficiently pure to be re-used in virgin RIM production.

8.2.2 Polyureas

Reaction of isocyanates with amine–ended soft block and hard block components instead of diols gives rise to polyureas (PURs). The urea linkage is the counterpart of the carbamate grouping of the polyurethanes. Generally, amines are more reactive towards isocyanates than alcohols, therefore RIM cycle times are shorter using amine systems, at around 1 min. Indeed, the formation of polyureas is so fast that it is extremely difficult to carry out these reactions in bulk unless RIM technology is used. In general, polyureas have good physical properties which are attributable to the extensive hydrogen bonding that can occur in the polymer, rather than the fact that these materials have a two-phase morphology.

PURs normally have better impact properties, improved water resistance and better heat resistance than PUs. Heat stability is particularly important when the materials used for automobile panels have to withstand paint stoving temperatures. They also offer better mould release than PUs and improved 'green strength' with consequent elimination of post-cure. Hence PURs are attractive for automated RIM.

8.2.3 Polydicyclopentadienes

Dicyclopentadiene (DCPD) polymerized using alkene metathesis catalysts, first reported in 1983, forms an extremely rigid, amorphous, cross-linked material, which is well suited to structural RIM (SRIM) applications when formed over pre-placed continuous fibre reinforcement. A commercially available glass-reinforced poly-DCPD ('Metton', Hercules Inc) is available. The poly-DCPDs are claimed to have excellent physical properties, particularly when reinforced with continuous glass fibres, and have begun to find limited commercial exploitation.

8.2.4 Epoxies

An important class of reactively processable polymers are epoxies, which are widely used in aerospace applications. Like the PUs there is an almost infinite variety of structural variations that can be made, which in turn impart significant differences in physical properties. In general, epoxies are brittle cross-linked materials and have been traditionally processed by RTM, but they can be combined with impact modifying reinforcement for RIM processing. Reports of exploratory epoxy RIM first appeared in 1980.

A number of factors render epoxies less amenable to RIM processing than PUs. Although the epoxy reaction is much more exothermic than the PU reaction it requires a greater mould temperature (around 140°C is preferred) to initiate polymerization, which leads to the possibility of extremely high temperatures (300°C) if rapid cycle times are required. However, the physical properties of epoxies makes them very attractive for structural applications, particularly when combined with high strength reinforcements such as Kevlar or carbon fibre. Epoxy RIM has great potential where cost is not a significant factor.

8.2.5 Binary polymer alloys

One method which has been adopted for the control of the properties of the matrix polymer in RIM has been the co-processing of two chemically distinct polymerizations, for example a condensation reaction (PU) and a radically initiated process (methyl methacrylate) demonstrated in our laboratories. These materials are often referred to in the literature as interpenetrating polymer networks (IPNs) or polymer blends. Binary polymer alloy (BPA) is probably a more appropriate descriptive name, since they are two-phase materials with some degree of inter-phase penetration.

The aim in co-processing is to obtain two cross-linked polymer systems which are mutually dispersed on a molecular level, but without chemical bonding between them (the definition of an IPN). The cross-linking is required to prevent de-mixing occurring on heating or with time. Without the distinct polymerization mechanisms, a graft polymer or, in the case of the same mechanism for both polymerizations, block co-polymers are formed. In practice, phase separation occurs during polymerization, to give the two-phase BPA structure.

In principle, this class of polymer can be considered for the possible production of engineering polymers. The physical properties obtained to date with acrylic/PU systems do not match those of the best RIM block copolymers (e.g. ARIMAX, subsection 8.2.7), polydicyclopentadienes or polyamides. The differences are greater when these materials are combined with continuous fibrous reinforcement. The best results claimed thus far for a BPA material have been for a polymer formed by very high pressure impingement mixing (1800 bar) of PU/polyester and PU/polystyrene polymer systems. However, it is questionable whether such a system could ever be commercially viable, as the high pressure would demand a rather special, high-cost machine. A possible advantage of co-processing two chemically independent systems is the reduced viscosity of the reagents obtained by adding vinyl monomers to PU precursors and the consequent higher level of particulate fillers that can be tolerated in them.

8.2.6 Polyamides

The basic concept of producing reactively processed polymers with a two-phase morphology is not confined to PUs and PURs. Another class of such materials are the reactively processed nylon block copolymers (NBC) based on nylon-6 hard block and a polyether soft block. These materials are semi-crystalline thermoplastics, and hence a detailed treatment is outside the scope of this book: they are included partly for completeness, and partly because their commercialization (NYRIM®, DSM Nyrim v.o.f.) has been the major reason for the development of commercial high-temperature RIM machinery

— reagent temperatures must be above the melting point of the major precursor, caprolactam. Typical stream temperatures of 80°C are used, with mould temperatures around 140°C. Polyamides have much better high-temperature physical properties than polyurethanes: their rigid, high-impact properties put the polyamides in the class of engineering plastics. However, high-temperature RIM processing equipment based upon the high pressure lance injection systems is expensive, since the reagent delivery systems and impingement mixhead must be heated above 70°C. A cheaper alternative is to use low-pressure equipment with static or dynamic mechanical mixers. The relatively low-viscosity of the reactants, of the order of 10 mPa s (water viscosity is around 1 m Pa s) compared with around 0.1 to 1 Pa s typical viscosities of PU reactants, means that they can be mixed easily, without the need for high-pressure impingement. Typical cycle times for automotive parts are around 3–5 min.

8.2.7 Acrylics

Most published information on acrylic-based RIM relates to a class of block co-polymers prepared by reactively processing, over pre-placed reinforcement, isocyanates and a monomer prepared from methacrylic acid, propylene oxide and maleic anhydride. They have been marketed under the trade name of 'Arimax' (Ashland Corp.). This composite material appears to offer some of the best properties of any reported RIM material obtained to date, and found limited commercial application. RTM of acrylics ('Modar', ICI plc) has also been exploited.

Not all RIM acrylic materials derive their physical properties from the two-phase character imparted by block co-polymerization. Acrylic-based materials have been produced by free radically initiated polymerization and the impact characteristics of these otherwise brittle materials improved by the incorporation of appropriately functionalized oligourethanes. The high-temperature modulus characteristics can be improved by the addition of divinyl monomers to the formulation. These materials probably have a two-phase morphology, but the domain size is such that it is difficult to observe experimentally.

8.2.8 Other RIM materials

Polyesters are very widely used in RTM processing, in the form of a pre-mixed single-reagent stream, and RIM of polyester systems is being explored, but little information is available in the public domain. RIM of silicones has been reported for small specialized electronic applications. Phenolic RIM resin properties have also been reported (Union Carbide), and a phenolic RIM (BP) interest arose in the 1980s for potential use in higher temperature automotive applications, but no such applications are yet apparent. Phenolic RIM reagents are particularly aggressive, requiring special steels for processing equipment.

8.2.9 Reinforcements

In the course of development of RIM, reinforcing fillers were incorporated into one or both reactant streams,[68] in order to enhance values of product flexural modulus (and hence heat sag resistance at elevated temperatures) and reduce the linear coefficient of thermal expansion in an attempt to satisfy automotive body panel requirements. Other fillers may be incorporated for special applications, such as barium sulphate for sound insulation, and various fire retardants. Typical reinforcing fillers investigated include hammer milled glass (the most used commercially), flake glass, chopped strand glass, wollastonite and mica.

Commercial hammer-milled glasses exhibit a broad fibre length distribution, from around tens of μm to 1 mm (rather lower than estimated critical fibre lengths), for fibres of diameter approximately 10–20 μm. Up to around 40 wt% loadings have been employed in PU RRIM. The fibres are normally coated with a silane size, to enhance the bond with the eventual polymer matrix; the size may also include dispersing agents to assist the fibres to distribute homogeneously in the polyol, seeking to avoid aggregation of fibres which are a cause of RIM mixhead orifice blockage. Chopped glass, now no longer commercially available, was more monodisperse, typically around 1.6 mm length for RIM, 10–20 μm diameter, and hence offered greater reinforcing efficiency, but was relatively expensive and even at low loadings caused rapid viscosity build-up at low shear rates. Flake glass platelets are typically up to 1 mm^2 in area, 10–20 μm thickness, and offer a degree of planar isotropy in moulded products.

Fibrous reinforcing fillers introduce problems in handling and health, settlement and machine wear, but provide a method for enhancing the Youngs modulus of PU products and decreasing the linear coefficient of thermal expansion to a value nearer to that of steel (which is important for steel–polymer car body construction).

However, the reinforcing efficiency of short fibre fillers tends to be low, particularly after attrition of the filler during high flow rate recirculation in standard RRIM processing equipment, and fibrous fillers lead to anisotropy of the product (much reduced in the case of platelet type fillers such as flake glass), which may be exploited in some cases but in most cases is not desirable. Consequently there has been growing interest in structural RIM (SRIM) using various forms of continuous reinforcing filler, for example, random glass mats, woven and non-woven glass, carbon fibre or ‘hybrid’ (mixture of fibres) cloths or knitted structures pre-placed in the mould cavity.

These provide a theoretically more efficient reinforcement due to the effectively infinite length of fibre. Packings up to ~80 wt% can be achieved, further increasing the stiffness of the product, and these fillers also lead to a more isotropic product. However, knitted structures in particular are not inherently mechanically stiff, hence reducing the achievable product stiffness. Also, at high loadings, the flow paths in a mould may demand relatively high pressure drops, leading to poor permeation of fibres, poor fibre wet-out, or loss of dimensional accuracy in the moulding through mould/press deflection or opening at high in-mould pressures. Technological problems relating to the manufacture, automated handling and placement of suitable reinforcement 'pre-forms' (or of automated lay-up of continuous fibres in situ), mould design and filling, and suitable chemical formulations are currently being addressed.

The SRIM technique shows great promise in achieving far higher property levels than by RRIM alone. It is also feasible that particulate fillers for reinforcement or specific property promotion will be required in the reactants in a combined RRIM/SRIM process.

8.3 RIM/SRIM injection processing technology

8.3.1 Dispensing systems

RIM/SRIM technology can be conveniently (but inadequately) divided into 'low pressure' and 'high pressure' injection capability,[67,71,73] and these categories can be subdivided into 'low temperature' and 'high temperature'.

Commercial PU RIM requires dispensing of two reagent streams at or near ambient temperatures and only relatively low temperature moulds (e.g. around 65°C) due to the inherent reactivity of the reagents. These factors provided the basis for all RIM technology. Low pressure (e.g. gear pump normally, or axial or radial piston pump) RIM equipment is widely available for casting or pouring PU foams. Figure 8.3 shows an example of commercial low-pressure foaming equipment. Higher pressure (e.g. lance positive displacement pump) equipment can be used for RIM but is more widely used where high-volume production or reinforced products are required. An example of a commercial high-pressure RRIM installation is shown in Fig. 8.4.

Two reagent stream machines predominate, but the range of polymers that can be reactively processed at acceptable rates can be extended by the use of more than two reagent streams and by heating the reagent delivery systems so that materials which are solids at ambient temperatures can be processed. Three stream machines are normally used for addition of a third minor stream, for example for colouring purposes. A few four stream RIM machines have been offered commercially, aimed at larger volume mouldings by

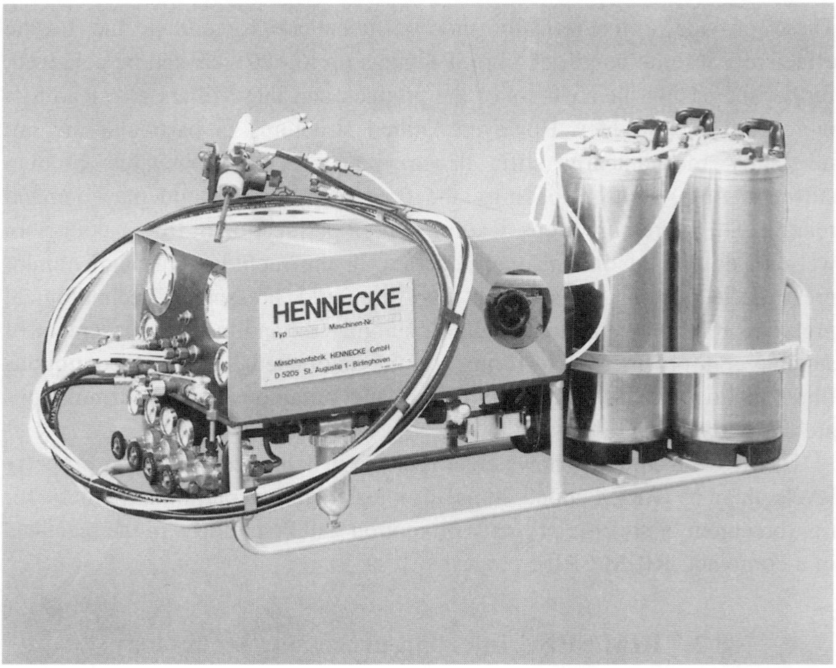

Figure 8.3 Commercial low-pressure RIM equipment. (Courtesy of Hennecke GmbH)

doubling the injection capacity of the two reagent streams. The use of three- and four-stream research machines have been reported, adding a new dimension to RRIM by allowing blends of different polymers to be processed.[83,84] These include the novel physically inter-locked interpenetrating polymer networks, which have synergistically enhanced physical properties (subsection 8.2.5).

Both low- and high-pressure machines which are capable of heating the entire reagent delivery system at temperatures above 80°C have been commercially available for a few years, and there are now several installations of both categories of machine, manufacturing commercial polyamide products.

High temperature RIM technology is essential to process caprolactam, and is also likely to be required for epoxy systems, amongst others. Where reagents must be processed at elevated temperatures, all the processing equipment, including the injection hoses, storage tanks, return lines and mixhead must be heated. Various means have been used to achieve this. The most common are the use of electrical trace heating and cartridge elements, and circulation of heating oils through insulated lines and jackets. The latter is preferable for controlled, uniform heating but is rather more expensive than

Figure 8.4 Commercial high-pressure RIM equipment. (Courtesy of ESU Cannon)

the already expensive electrical method. A 'hot RIM' machine may cost twice the price of its ambient temperature equivalent.

In most processing applications, the chemical reagents are formulated in such a way that injection involves two lances travelling at almost identical velocities (i.e. a stoichiometry near 1:1). Low-pressure machines (e.g. for coating or spraying) are generally less accurate in controlling this dispense ratio throughout the injection, and, if using gear pumps, are susceptible to gas entrainment in reagents and leakage flows. For the more accurate lance type machines, in the case of dispense ratios around 1:1, both injector systems have very similar dynamics and the injection ratio is near to the ideal, even during the period of lance acceleration and mixhead opening. This is not the case when substantially different reagent ratios have to be used, because lance dynamics have been found to be velocity dependent. In this case, or in the case of non-optimized high flow rate recirculation prior to injection, 'lead-lag' (off ratio production) can be particularly problematic for RIM polymerizations which are particularly ratio sensitive, such as PUs and PURs.

The various RRIM technologies illustrated schematically in Fig. 8.1 are all designed for batch processing. However, RIM technology is also used in a continuous manner with polyurethanes for surface coatings applications either with or without reinforcing fillers. PU spraying techniques are widely used and involve impingement mixing of the reagents in the spray-head.

8.3.2 Moulds and presses

One of the key initial attractions of RIM was the relatively low pressures after the mixing head. Typical in-mould pressures were around 1 bar, due to the relatively low viscosities of the reactants. In turn this led to moderate clamping forces, so that large area products could be moulded on low-locking-force presses — for example a RIM product of the order of 1 m^2 area could be obtained using a 20 tonne locking force press. Injection of tractor products of the order of 30 kg mass have been reported: RIM is well suited to the manufacture of large-area, large-mass products.

A further spin-off of low in-mould pressures was that moulds could be made from a range of materials which provided easier and more rapid fabrication than precision machining of tool steels. Epoxy resin, sprayed metal, electro-deposited nickel shell, and even plywood and silicone rubber tools have been used either for prototyping or for low-volume production. Rapid prototyping was perceived to be a particular advantage of the RIM process. Aluminium, mild steel and tool steel have been employed for higher volume production; however, tool steel is still favoured for top quality products.[67,71,82]

The incorporation of fillers, either as particulates in the reactants or as preforms in the mould, invariably leads to higher injection pressures being required. At high levels of reinforcement it would not be uncommon to observe in-mould pressures greater than 20 bar; the original RIM presses then prove to be inadequate in terms of locking force and stiffness, with mould opening, flashing or unacceptable mould deflections across the cavity. Consequently, a move to more substantial, higher locking force presses (e.g. 80 to several hundred tonnes) followed the growth of interest in RRIM and SRIM. These processes are therefore best seen not as very low pressure RIM operations but rather as medium pressure processes. Even so, in-mould pressures are still low compared with conventional injection moulding.

A wide variety of press designs is available for reactive injection processes,[67,71] from simple vertical opening to 'booking' presses where the platens open at an angle, rather like a hinge, for ease of product removal. Robotic devices have also found increasing use in spraying of mould release prior to injection, and in product removal and deflashing.

There is also a considerable variety of manufacturing systems associated with RIM processes which are characterized in simple terms by the number of moulds serviced by one dispensing machine. Automotive panel or similar product manufacture will typically have one press dedicated to one dispensing station, whereas cheaper mould, moderate volume production (e.g. for shoe soles or steering wheels) may have a carousel arrangement of moulds, with tens of moulds per dispensing machine.

8.3.3 Rheology of RIM formulations

A notable feature of RIM formulations is the relatively low viscosities of the reactants and the reacting mixture before significant reaction occurs.[73] The viscosity range has a major influence on processing technology. Polyamide and acrylic RIM reactants may have apparent viscosities as low as 10^{-3} Pa s, PU reactants up to approximately 0.5 Pa s. Addition of reinforcing filler to the reactants causes a rise in this initial viscosity, so much so that at the low shear rates encountered in stirred storage tanks it is not possible to incorporate more than ~40 wt% of hammer-milled glass or ~16 wt% of chopped strand glass in PU reactants (air inclusion also becoming a significant problem at these loadings). The behaviour of PU reactants, even containing chopped strand glass reinforcing filler, has been observed to be essentially newtonian over the range of shear rate expected in RIM mixheads ($\sim 10^4\ s^{-1}$ to $\sim 10^6\ s^{-1}$). As reaction and mould filling proceed there may be an initial fall in viscosity of the reacting mixture due to heat transfer from the mould and exothermic heat.[73,81]

It is desirable to fill a mould completely before the occurrence of the increasingly rapid rise in viscosity due to polymerization. Simple modelling combining reaction kinetics with structural relationships can be used to predict the build-up in molecular weight, and hence rise in viscosity, with time and temperature in the mould. The true chemo-rheology of polymerizing mixtures in reactive injection is particularly complex, with the superposition of heat transfer, polymerization, phase separation and cross-linking, all possibly in the presence of reinforcing fillers in suspension, or flow through reinforcing fibre mats (such fillers may act as heat sinks and barriers to diffusion of the reacting species).

An empirical equation has been proposed for viscosity change by Castro and co-workers:[73]

$$\eta = \eta_0(T)\left[\frac{\alpha_g}{\alpha_g - \alpha}\right]^{c_1+c_2\alpha}$$

where initial viscosity $\eta_0(T) = A \exp(E/RT)$, η is apparent viscosity, α is the conversion of isocyanate groups, α_g is the conversion of isocyanate groups at physical gelation due to the formation of a network of hard segment domains, and is the critical parameter for fitting the viscosity rise with polymerization. Laboratory measurements are unrealistic for very fast reacting systems, such as those in use commercially. Hence, attempting to measure shear viscosity in the mould has considerable significance. It is claimed possible to use the initial pressure rise during mould filling to predict initial viscosity, but in-mould rheological measurements involving measured pressure drops in a known geometry, together with measurement of local fluid temperature, are to be preferred (but are clearly not very practical for commercial production).

8.3.4 Machine control

In the case of RIM of PUs or related formulations accurate control of dispensing at the stoichiometric ratio is necessary to achieve the desired product properties. A suitable measure of machine dispense accuracy is the root mean square (rms) deviation of reactant ratio from the set ratio.[79,80,83,84] However, this may typically be up to 5% of the set ratio on commercial dispensing equipment, hence commercial RIM formulations often have to be designed to provide a tolerance in ratio. Inadequate control of dispense ratio during injection is likely to lead to inhomogeneous products, either through variation in composition (causing unreacted regions in the worst case) or variation in mixing conditions. Lance displacement pumps provide better dispense control than gear pumps.

The sequence of events in a typical commercial RIM cycle typically involves a time of high-pressure recirculation of reactants through the mixhead and back to tank, prior to opening of the mixhead and injection into the mould. The effects of poor sequence control at the mixhead, for example delayed or slow opening of the mixhead, will cause uncontrolled compressible flows of reactants.[79,80] This is particularly the case at the start of injection, and leads to the well-known phenomenon of 'lead-lag', that is uncontrolled off-ratio production at the commencement of injection, which can result in substandard products, or undesirable variations in product properties.

Although accurate ratio control is less critical for non-PU RIM systems, accurate dispensing is a prerequisite for consistent production and properties.

8.3.5 Mixing

Commercial mixheads were designed essentially empirically, and act as directional flow valves for the reactants (Fig. 8.5): when the mixhead is actuated reactants are diverted from flowing back to their storage tanks to flowing into the small volume (of the order of millilitres) mixing chamber through small diameter orifices, typically at exit velocities of around $100 \, m \, s^{-1}$. The orifices may be adjustable (by means of flow restrictors), predominantly for the purpose of balancing reactant stream pressures. Two orifices are generally used, but larger mixheads can have three or four inlets, allowing multicomponent mixing. In commercial practice this allows the incorporation of colourants or foaming agents. Multi-component injection has also been used in a novel manner by the authors in a three-stream dispensing of PU precursors, effectively blending and mixing in one stage at the mixhead, and in simultaneous injection of four streams to produce a polymer blend in a single mould cavity.

Good mixing of reactants is critical to the completion of fast urethane polymerization. High-pressure jet impingement has become the dominant route to the required intensive mixing[75–80] for PUs (essentially the rapid creation of adequate contact area between the reactants in a time scale which is short compared to total injection time). Since the technology for

Figure 8.5 A commercial two component high-pressure mixhead. (Courtesy of Krauss Maffei Kunststofftechnik GmbH)

impingement mixheads with self-cleaning valving is attractive from a manufacturing viewpoint, impingement mixing is also the major mixing method for non-PU RIM. Low-pressure RIM machinery tends to employ static or dynamic mechanical mixers which require solvent flushing.

The assessment of mixing quality has been the subject of much study and some controversy. The lowest value of Reynolds number, Re,

$$\mathrm{Re} = \frac{\rho v d}{\eta}$$

(where ρ = reactant density, v = average stream velocity, d = orifice minimum opening, η = reactant apparent viscosity) in an impingement mixing nozzle has become a favoured criterion for mixing quality. In the case of PU RIM, the lowest values of Re will be obtained for the polyol blend, which normally exhibits apparent viscosities around four times that of the isocyanate reactant.

There have been many attempts to quantify the degree of mixing. Early work[75] on non-reacting systems indicated that Re >~200 was required to obtain 'good mixing'. Many further studies, including those on reacting systems, do not greatly contradict this 'rule of thumb'.

In practice, impingement mixing is thought to create layers of fluid, 'lamellae', probably with a distribution of thicknesses. Where the lamellae are thin, the polymerization reaction can proceed by diffusion of reactants into each other; where lamellae are thicker, unreacted material is likely to be trapped in the lamellae. Thus efficient mixing would create sufficiently thin lamellae to allow diffusion to complete the mixing process; this would lead to a higher conversion.

Flow visualization experiments[78] clearly indicate a larger scale striation than the ~1 μm which would be needed for molecular diffusion to allow completion of the polymerization reaction in a striation. However, the RIM process does produce well-reacted polymers, consequently the 'fluid mechanics' approaches above cannot represent a full understanding of mixing processes for reacting systems. Various workers tentatively suggest a micromixing mechanism related to surface effects: the interface between two reactants may become unstable and droplets of the order of 1 μm diameter may be formed, each drop encapsulated in a newly formed urethane structure, which then disperse into one of the reactant lamellae. Molecular diffusion could then account for completion of the reaction. Such micromixing, with diffusion, would thus be the final stage of an intensive mixing process in which reactant jets of the order of 2 mm diameter are impinged to produce lamellae of thickness ~100 μm, a process which requires critical values of Re of around 300 at the mixhead orifice — a condition which can be readily achieved in RIM dispensing equipment.

8.3.6 *Mould filling*

Generally, low-viscosity mixtures passing into a mould imply that fill pressures will be low, and that wetting of the surface (to pick up surface detail) will be good. However, the danger of air entrainment at the flow front is also increased. A major processing requirement is to fill moulds in a laminar flow mode, but in times somewhat shorter than the gelation time for the reacting system.

Both experimental and modelling studies[73,81,82] have indicated a parabolic velocity profile in the major part of the advancing fluid, with a 'fountain flow' at the flow front (as in conventional injection moulding). If such a flow genuinely occurs in a reacting system, it is important that it be quantified since it determines the location of each element of reacting mixture in the mould.

Flow front instabilities begin to occur when too rapid a mould filling is attempted. At typical PU RIM mould filling viscosities, it is suggested that a critical Reynolds number of between 3 and 30 in the cavity should not be exceeded to avoid instabilities leading to air entrainment, gate design also being important in this matter.

It is therefore clearly of practical importance that the mould be filled in a steady laminar manner, at Re below some critical value, before gelation occurs or before any significant rise in viscosity causes too great a demand in filling pressure. A simple criterion for mould filling has been suggested: if the time to fill a mould is greater than the *adiabatic* fill time then the mould will not be filled. Some workers have proposed the concept of a RIM 'mouldability diagram' (i.e. a designation of the conditions under which a moulding can be successfully made), similar to those used in conventional injection moulding.

The issue of mould filling in the presence of continuous fillers preplaced in the mould is another issue of considerable practical importance, given the level of interest in RTM and SRIM processing. Typical analyses for flow through glass mats employ D'Arcy's law, which models flow in permeable media. A major problem to modellers is the accurate determination of permeabilities; a major related practical problem is to ensure wetting of the whole surface area of reinforcing fillers (especially when made from tows of fibres) during moulding.

8.3.7 *Curing*

Curing is often the rate-limiting step in the RIM moulding cycle. Sufficient 'green strength' must be built during cure for demoulding. In rapid reacting RIM systems the in-mould time is generally fixed by heat transfer limitations, i.e. the time taken to conduct away exothermic heat to the mould. The cure and temperature profiles in the moulding will determine the development of properties and hence strongly influence demoulding and product quality.

The Dampkohler number, Da, is an important engineering parameter in assessing cure times.[73] For a mould of typical thickness H:

$$\mathrm{Da} = \frac{\text{heat production by chemical reaction}}{\text{heat transport by conduction}} = \frac{C_O k(T_O)}{4a/H^2}$$

where $k(T_0)$ is the overall rate constant evaluated at initial material temperature, and a = thermal diffusivity of the reacting mixture. Large values of Da clearly suggest that adiabatic conditions apply; when Da is small, reaction heat may be neglected and the problem reduces to one of transient heat transfer through the mould thickness.

8.4 Properties

The final properties of a RIM polymer matrix will depend on formulation, variables and the process history of all the elements making up the final product. The occurrence and degree of cross-linking, phase separation and alloying will have large effects. A wide range of physical properties may therefore be achieved in a controlled manner in a RIM polymer matrix by choice of formulation (e.g. amount and type of chain extender in a PU formulation) and there will also possibly be further less controlled variation through process history. An indication of some physical properties of RIM polymers is given in Table 8.2.

A strong driving factor in the development of reactive injection processes has been the desire to obtain structural parts, that is those with adequate levels of stiffness to support practical loadings. The amount of enhancement in mechanical properties which can be achieved by manipulation of chemistry and process history is limited, compared with the demands for structural materials. Incorporation of reinforcing fillers[68] of the type

Table 8.2 Typical properties of unreinforced RIM systems (approximate, for comparative purposes only)

Property	Method (ASTM no)	PU (high modulus)	Polyamide Nyrim 2000	DCP Metton	Epoxy
Flexural modulus (MPa)	D790	830	1400	1860	2900
E-29C/E70C		~5	4.2	1.6	1.7
Tensile strength (MPa)	D638	28	44	34	69
Izod impact (J/m)	D256	240	300	430	24
Heat distortion T °C	D648	80	175	95	118
Thermal expansion m/m°C × 10^6	D696	150	122	54	56
Specific gravity		1.00	1.13	1.04	

After C Macosko, *Fundamentals of RIM*, Hanser, 1989, New York.

Table 8.3 Selected RRIM PU properties

Reinforcing agent used	Product density ($g\,cm^{-3}$)	Product hardness (Shore D)	Heat sag[a] (mm) ‖/ ⊥[b]	Impact energy (J)	Tensile strength (MPa) ‖/ ⊥[b]	Flexural modulus (MPa) ‖/ ⊥[b]
Unfilled	1.1	58.0	35	55	20	300
5% CSG	1.18	59.5	n/a	54.2	24/24	576/390
5% CSG + 5% Woll	1.15	60.5	15/28	40.3	21/20	525/375
5% CSG + 10% Woll	1.17	61.0	20/28	37.2	19/19	470/415
5% CSG + 5% HMG[c]	1.23	60.0	n/a	36.1	25/25	555/380
5% Flake	1.2	60.0	32/40	21.0	17/15	490/490
5% Flake + 10% HMG[c]	1.23	62.5	22/33	21.2	23/21	535/470
2% Flake + 4% CSG	1.13	60.0	10/26	24.5	23/20	730/455
4% Flake + 3% CSG	1.2	62.0	10/25	24.5	20/19	690/510
4% Flake + 8% HMG + 3% CSG	1.18	63.0	9/20	24.4	21/19	740/530

[a] 125 mm overhang at 120 °C for 1 h [b] parallel/perpendicular to the injection direction [c] hammer-milled glass
Selected results from a polyurethane RRIM program with mixed reinforcing fillers (ambient temperature testing): CSG, (Pilkington Brothers) WX6450 1.5 mm chopped-strand glass; Woll, Wollastonite (Colin McNeal Ltd) G-Wollastikup 1100/4 10 u; Flake, (Owens Corning) 3 mm flake glass 74 × 59639 no coupling agent

discussed in sub-section 8.2.9 appears to be the only satisfactory way of attaining relatively high-modulus materials with, for example, adequate impact properties and/or reduced coefficient of thermal expansion. Indeed, addition of only moderate levels of particulate reinforcing fillers has a marked effect on modulus but leads to a decrease in elongation to break and impact strength. Mixed reinforcing fillers have been used to modify the often unwanted anisotropy of mechanical properties caused by flow-aligned fibres. Table 8.3 shows some selected results. The effect of a two-dimensional platelet filler, flake glass, in reducing anisotropy is clear.

By far the most significant enhancement of modulus and improvement in other physical properties such as impact strength, high temperature creep behaviour and linear thermal expansion coefficient is obtained by SRIM (or resin injection) over continuous mats, hence the increasing interest in this process. Table 8.4 shows the large increases obtained in flexural modulus, to levels which may be considered structural, by incorporation of high weight percentages of continuous glass mats. Figure 8.6 shows flexural storage modulus, E', measured at 0.3 Hz, as a function of temperature for a PU produced by RIM and the same polymer containing mixed discrete glass reinforcement, produced by RRIM, with two examples of the same polymer containing ~20 wt% and ~40 wt% of continuous random glass mat, produced by RIM through these mats.

Although the continuous reinforcement clearly has a major effect in determining the overall modulus level, the polymer matrix temperature performance cannot be ignored. From a structural engineering viewpoint it is

Table 8.4 Some comparative properties for SRIM products (ambient temperature properties except heat distortion temperature)

Material	Tensile strength (MPa)	Elongation (%)	Flexural modulus (GPa)	Notched Izod impact ($J m^{-1}$)	Heat distortion temperature (°C)
Mild steel	700		208		
Aluminium	150		70		
Acrylic unreinforced (Modar, ICI)	56	6.5	2.86		83
Epoxy/ 30% vol C fibre	419		28		
ARIMAX/ 32 wt% C fibre	250		21	609	
ARIMAX/ 50 wt% glass fibre	276	2.5	14	939	210
RIM PU/ 40 wt% glass fibre			1.44		
NYRIM2000 /72 wt% glass	280		16		

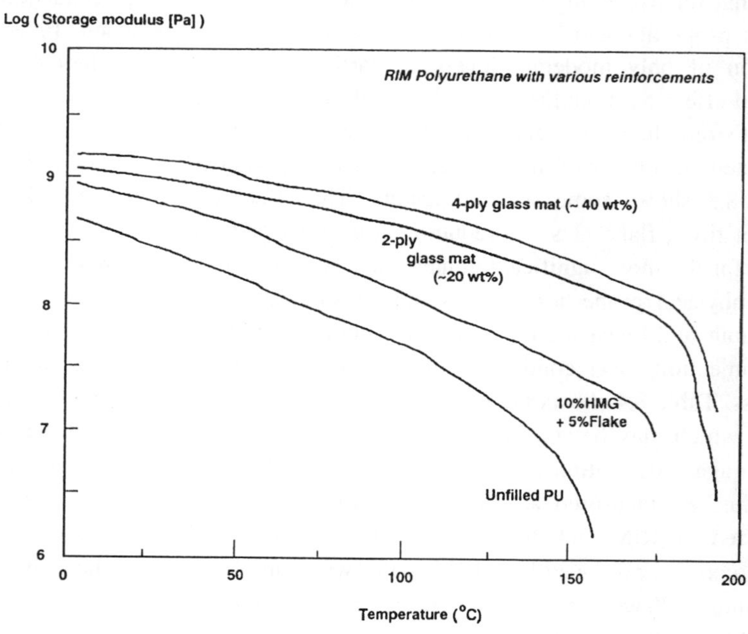

Figure 8.6 Variations of flexural storage modulus with temperature for RIM, RRIM and SRIM polyurethane products.

preferable to have as flat a modulus–temperature response as possible. This suggests that there is room for manipulation of the polymer matrix to reduce the fall in composite properties with temperature, whilst maintaining adequate impact properties. In this context, SRIM of polyamides and SRIM of alloys or IPNs would seem to offer attractive routes to the desired property levels for structural products.

Current RRIM/SRIM products include automotive bumpers, bumper beams, undershields, floor pans, doors and chassis components. RIM and associated reactive injection processes and products are consequently challenging traditional sheet metal forming processes and products for automotive applications, but are themselves in competition with other polymer forming techniques, in particular with the well-established large scale injection moulding of thermoplastics blends and alloys for the same automotive applications, as well as furniture, appliance housings and leisure goods. The potential for development of structural materials for high volume applications remains a strong incentive for developments in RIM, and in particular SRIM.

8.5 Developments in reactive injection processing

There is little doubt that RIM is now a mature processing technology for polyurethanes and polyureas and the capability exists to process a wide range of materials. It is also apparent that the simpler RTM process has been most favoured in recent automotive-related developments, although it is now recognized that pre-mixed RTM thermally driven single stream formulations are too restrictive in terms of tailored properties and production rate. For RIM technology to be used with polyamides, acrylics and other materials, the decisions are likely to involve economics, product physical properties and convenience, in addition to issues relating to the capability of the processing machinery or any limitations in the chemistry. For example, there is clear evidence that it is possible to reactively process reinforced polyamides for automotive body applications, but despite the good engineering properties of this material, reinforced polyurethanes and polyureas are still preferred (where a RIM product is chosen).

In 1987, RIM market volume was around 100 000 tonnes worldwide; almost all of this went into automotive products, and of these only 1.5% were produced by the then relatively new SRIM process. Initially, RIM technology and materials for structural applications were in direct competition with metal forming processes for thin-section, large surface area products, but strong competition exists from alternative polymer-processing methods for the same type of product. For example, there have been significant developments in the field of polymer blends, which has given rise to grades of materials which can be successfully injection moulded, possibly together with reinforcing fillers, into large surface area mould cavities. A polymer-processing industry well versed in the techniques of injection moulding appears to readily accept the need for larger injection moulding machines and mould clamps, rather than introducing wet chemistry into the factory.

However, there is still likely to be growth in reactive processing techniques, partly because of the ability to manufacture continuous fibre reinforced composites and partly because of the capability of tailoring products.

A major advantage which reactive injection technologies have over injection moulding is that they can be used to encapsulate a pre-form with a polymer matrix (i.e. SRIM). The preparation and handling of pre-forms is developing rapidly and in principle at least the capability exists for robotic handling of pre-forms, which could make reaction injection moulding over computer designed reinforcement commercially feasible. If RIM is to develop in new directions, then the production of complex geometry components with designed reinforcement for structural applications would appear to be one of the more obvious routes.

Tailoring of final product properties by the processor provides exciting prospects for flexibility in manufacture.[83,84] Two such routes have been investigated in our laboratories, using our unique computer controlled and monitored four-reactant stream RIM/SRIM facility, 'MultiRIM' (see Fig. 8.7). The first case involves the alloying of two polymers during reactive processing — exploiting the four-stream injection capability of our equipment, using a commercial (Krauss Maffei) four-stream mixing head. The second case, currently under investigation, involves perturbation of ratio during injection, to manipulate product properties with location in the mould. Furthermore, the effort which has been devoted to the development of polymerization processes with controlled but fast reaction dynamics for RIM and RRIM machines can also, in many instances, be translated to alternative processing technologies such as continuous reactive extrusion using twin-screw extruders.

A further advantage of reactive injection technologies which should encourage growth is integration of manufacture. Automotive components such as seating or whole inner door panel structures are now being made by reactive processing techniques. Also, the relatively low viscosity reacting mixture allows metal or other inserts to be moulded in easily.

PU and PUR matrix materials, with their considerable potential for tailoring, are currently maintaining their lead over other RIM systems, but competition — now well beyond the laboratory stage — is available and, in some instances, satisfying property requirements beyond the urethanes and ureas. There is little doubt that current RRIM/SRIM technologies are the forerunners of a range of fast reactive processing technologies for the production of high-performance engineering parts.

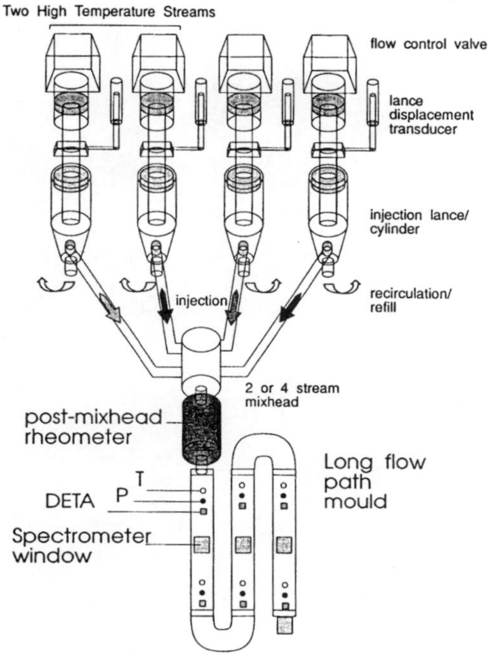

Figure 8.7 Schematic diagram of Bradford 'Multirim' four-stream computer-controlled reactive processing facility, with instrumented mould.

Acknowledgements

The authors gratefully acknowledge the support of EPSRC and many companies in funding our reactive processing research programmes; also the dedicated work of a substantial research team of chemists, engineers and technicians.

CHAPTER 9

Secondary operations

Section 9.5 Adhesive Bonding by W A Lees, Permabond Adhesives Ltd.

9.1 Introduction

This chapter provides a summary of the main processes that may be required to be carried out on the component after removal from the mould to bring it to its finished, and ready to be used, state.

Bearing in mind the potential diversity of processes possible, the information and comments provided here are in the form of general guidelines, which have been subdivided into four main groups:

1. trimming and machining operations;
2. mechanical attachments;
3. preparations for painting;
4. adhesive bonding.

Before entering into a description of the different processes and techniques that can be applied, it is essential to strongly emphasize the importance of giving due consideration to the requirements for these secondary operations during the design phase of the project. This consideration must be exercised in three ways:

- correct choice of materials for the product, plus knowledge of the material behaviour regarding the required secondary operations;
- the detailed design features necessary to optimize the secondary operation process and the quality of the finished result;
- the cost implications of the required process(es).

For example, the location of trim edges, or required mould split lines in visible surface areas of an aesthetically important component, such as a vehicle exterior body panel or a domestic appliance housing, would inevitably lead to quality problems (and costs) associated with the secondary operation. The type and construction of the mould can also have a very significant effect on the amount, position, and difficulty of removal of moulding flash.

Basic recommendations regarding good component design practice, and the characteristic features of the main mould types, can be found elsewhere in this volume, and should be consulted in conjunction with the specific comments given in this chapter.

9.2 Trimming and machining operations

9.2.1 Trimming

Trimming, often referred to as deflashing, is a commonly required operation with thermoset mouldings resulting from the tendency of these materials to flash the mould during the phase of the curing in which the resin viscosity can be very low.

For large components, such as SMC vehicle body panels, the trimming is usually performed by hand near the mould station, using a powered trimming tool. It is easier to remove the flash before the component cools and achieves full strength, but handling in this green condition therefore can also bring some risk of component distortion. In such cases it is necessary to provide suitable fixtures to support the part during this operation, which can also serve as cooling fixtures to minimize distortion and warpage. The quality of the trimmed edges can be affected by the sharpness of the trimming tool, the temperature of the part, the thickness of the flash, plus the geometry of the component itself.

For smaller components manufactured in large volume, it would be normal to deflash the components in bulk in a purpose designed deflashing machine. The most common are a tumbler type using a suitable deflashing medium to assist the removal of the flash from the component.

For smaller components manufactured in low quantity, it would be normal for the press operator to deflash the parts by hand with suitable hand implements at the press side.

9.2.2 Machining of holes and openings

Drilling/machining of plain and/or threaded holes in thermosetting mouldings is a subject requiring particular attention since problems of drill/cutter wear can be significant when the moulding contains glass or mineral reinforcements. Furthermore, the intrinsic brittleness of many thermosetting resins can also lead to risks of cracking, chipping or breakage of components during these operations. As a result, the specification of hole size and tolerance is very important and is often based, not only on materials manufacturer's recommendations, but also on past experience and practical testing.

Holes and openings can be formed by using a drill, punch, router or high-pressure water jet.

Drills are used for round holes, while punches may sometimes be employed for round, rectangular and irregularly shaped holes. Guides and fixtures are required to hold the part and the cutting tools, the design and operation of the fixtures being critical in determining the tolerance capability.

Care must be exercised in allowing sufficient edge distance between the hole and the perimeter of the component to maintain strength and avoid breakage (usually at least the hole diameter).

Routers can be used instead of drills for large, irregular holes or for edge trimming. It is becoming increasingly common to carry out the routing operation in a computer numerical control (CNC) station for greater precision, flexibility and automation. Water jet cutting has also become more popular for trimming and finishing larger thermoset panel type components. The process operates using a stream of water which is forced through an orifice at very high pressure (around 300–400 MPa) to produce the cutting jet. The process lends itself to CNC control and is very satisfactory for high precision straight or profiled cuts. It is particularly advantageous in reducing airborne dust particles and in giving a high quality to the cut edge. Some work has been done to introduce abrasive materials into the water jet to increase cut rates and reduce power requirements.[85]

An alternative to moulded openings in a panel-type moulding is to reduce the thickness of the moulding in the required opening to create a thin membrane which can subsequently be removed by normal secondary operations (Fig. 9.1). The benefits of this approach can include the promotion of more homogeneous material flow in the mould, and some improvement in the mechanical properties of areas around the hole by the reduction or elimination of knit lines.

9.3 Mechanical attachments

The design requirements for threaded fasteners, tapping or stud plates, through holes for fasteners and rivets, etc., are crucially important in ensuring a mechanically sound solution, and in avoiding damage and breakages during

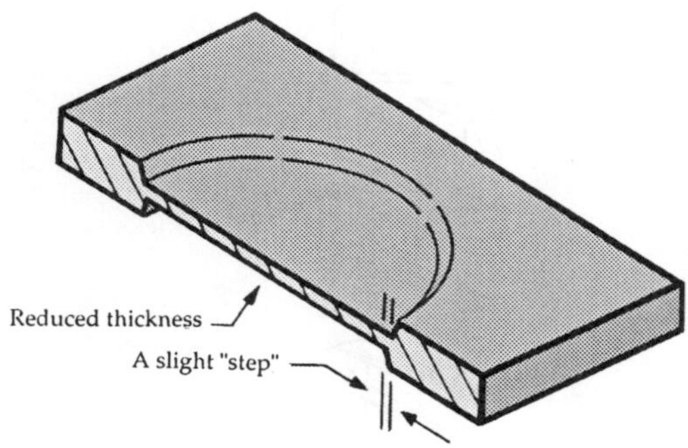

Figure 9.1 Moulded diaphragm to promote homogeneous material flow through a hole location.

manufacture. For preliminary indications of good design practice, the reader should return to the design guidelines in Chapter 1, and if necessary turn to a more detailed design manual.

9.3.1 Threaded fasteners

It is common practice to mould bosses into the thermoset component, which incorporate moulded or drilled holes to accept a threaded fastener. This approach is generally recommended for fastenings which will only be disassembled infrequently. As mentioned previously, the dimensions of the hole have to be carefully selected to maximize strength, torque retention and controlled driving torque for the fastener.

Thread cutting screws are not normally recommended for repetitive torquing and removal. In recent years the use of double-lead threaded fasteners (often called 'Hi-Lo' fasteners) have become widely used (Fig. 9.2). One thread has a large diameter with a sharp included angle, the other thread has a smaller diameter and shallow inclined angle. The combination allows cutting of the thread with minimum stress on the material. As a general rule bosses should be designed 25% longer than the fastener to allow room for the chips created by thread cutting. Consultation with the fastener manufacturer is strongly recommended.

9.3.2 Threaded inserts

For components requiring repetitive torquing or removal, it is normally recommended to have a metallic threaded insert installed in the boss. This is traditionally done by placing the insert in the mould and then moulding in place. Such inserts have a knurled exterior diameter to resist turning.

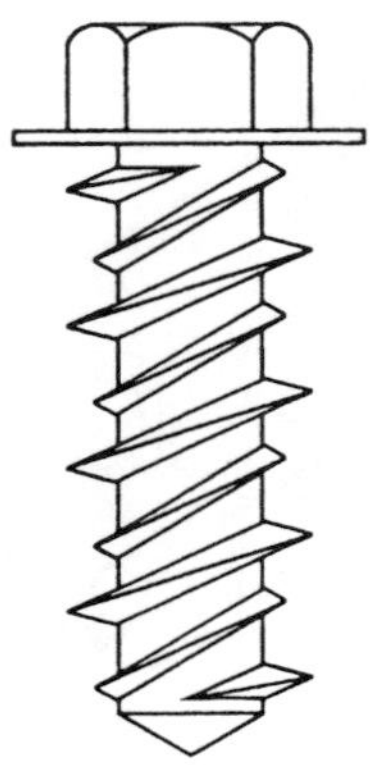

Figure 9.2 'HiLo'-type screw with double lead cutting thread.

Figure 9.3 Examples of self-tapping metal inserts for plastics. (Courtesy of Tappex Threaded Inserts Ltd.)

Alternatively, the inserts may be screwed into a pre-moulded hole in the boss. A recently developed family of self-tapping inserts, having three equidistant main and three secondary cutting flutes (Fig. 9.3), are proving to be particularly effective for post-moulding insertion into thermosets.

9.3.3 Studs and plates

An alternative to using this approach is to use a double-ended stud with a removable nut. If this is not possible, then tapping plates or stud plates may be considered (Fig. 9.4). The tapping plate consists of a flat or formed (metal) plate with threads for accepting fasteners. Stud plates are similar in principle, but use threaded studs instead of nuts. Tapping plates and stud plates have the benefit of spreading the load over a larger area, and are recommended when a boss-retained fastener would offer insufficient pull-out strength.

9.3.4 Through holes

In other situations through holes may be preferred, with the rules for distance of hole from the component edge mentioned in the previous section still applicable.

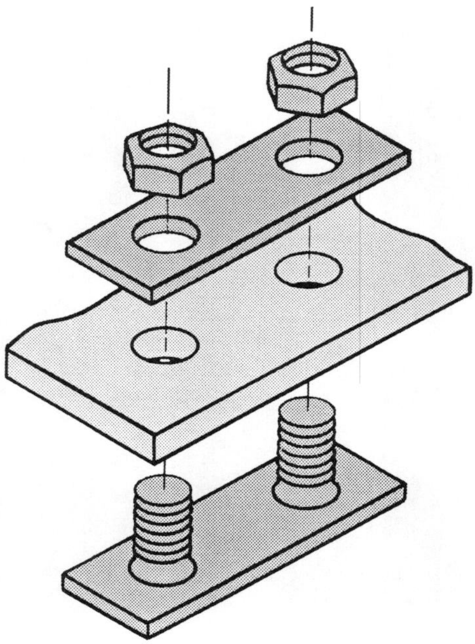

Figure 9.4 Tapping or stud plate for distributing loads over a greater area.

9.3.5 Rivets

The use of through holes with rivets is also a common solution. In this case the hole diameter and component thickness have to be carefully specified to avoid damage during the riveting operation and to be mechanically sound for product performance. When the rivet reacts against the thermoset component, a back-up washer is usually recommended to minimize the possibility of cracking.

9.3.6 Alternative fasteners

U-nuts and J-nuts can be easily incorporated in panels, with associated back-up holes to allow clearance for the threaded fastener (Fig. 9.5).

In addition there are now many different alternative proprietary fasteners in the market place, such as the so-called 'Christmas tree', and other press-in or snap fit systems.

9.4 Requirements for painted components

Many large thermosetting moulded components, such as SMC vehicle exterior panels, require paint finishing. A full discusion of the materials,

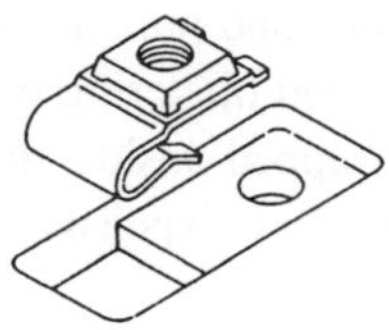

Figure 9.5 Clearance holes for U-nuts and J-nuts can be moulded into the panel.

processes, process plant and technology of paint finishing goes well beyond the scope of this book. It is, however, considered useful to provide some background to the needs and approaches necessary. The following section is written around the requirements for painted SMC vehicle body panels, but should give the reader a clear idea of the main issues involved.

9.4.1 Paint line requirements for SMC panels

SMC vehicle panels are typically primed and top coated with paint systems that require elevated temperature curing to reach a high-gloss finish (paint oven temperatures ranging from 75–160°C). Panels subjected to electrophoretic priming may reach 200°C.

The most critical temperature zone for the SMC component is the first phase in which the temperature ramps up from ambient to the paint-cure temperature. In this phase the component can be subjected to an unbalanced temperature profile, by virtue of the plant temperature profile and the geometry and varying thickness of the part. This variation can influence the paint-cure rate and lead to visible surface defects (sometimes called ghosting or shadowing).

When the part has reached the upper temperature limit imposed by the paint plant it will be significantly less stiff than at room temperature, and therefore subject to distortion and/or warpage. The method and positioning of the panel supports is therefore critical. Finally, in the cooling phase, care has to be exercised in the jigging and handling to avoid further risks of distortion.

An important consideration during the whole operation leading from the moulding of the panel through the painting process, is the design and use of specialist jigs to simplify component handling through the complete process sequence.

Even with this brief description it can be seen that the process is very critical and requires the close attention of a knowledgeable process engineer to establish the most suitable plant conditions.

9.4.2 Alternative painting systems

One of the traditional quality problems with painting of the SMC vehicle body panels is the question of pits and porosity in the surface of the part that can cause unacceptable bubbles and blistering of the paint surface, caused by the gases and paint solvents trapped inside the porous voids. Repair and refinishing of the panels can be very expensive and labour intensive.

9.4.2.1 In-mold coating

The process of in-mold coating (IMC) was developed to address this problem. The process, as applied to compression moulded SMC panels can be described as follows.

At the point that an SMC panel moulding is sufficiently cured in the mould during the moulding cycle, the moulding (clamping) pressure is reduced sufficiently to allow the high-pressure injection of a specially formulated lacquer through an injection nozzle so that the lacquer can flow across the surface of the component, filling the surface holes and pores, expelling gases and trapped volatiles in its path, and finally curing to replicate the high-quality surface of the mould. In some cases the mould is actually opened a small amount to aid the lacquer flow and escape of trapped gases. After the lacquer has entered the mould, the mould-clamping pressure is reapplied until final curing is complete.

This process works best with large and essentially flat components. Vertical surfaces and long thin sections are very difficult to completely cover with the lacquer. It is obviously important to place the injection nozzle in a position which will ensure the lacquer preferentially covers the aesthetic surfaces of the component, and does not leave a visible witness of the injection position in an obvious location. A more detailed description of the IMC process can be found in section 7.4.

9.4.2.2 Vacuum-assisted moulding

An alternative approach to reduce the problem of air blisters and porosity in the mouldings to be painted is to employ a vacuum-assisted moulding system. This utilizes a vacuum chamber which is created around the mould cavity during closure and immediately prior to the mould punch contacting the charge of moulding material. In this way the mould cavity can be evacuated, and the material can fill the cavity without gas entrapment problems.

9.4.2.3 Surface pre-treatments

Pre-treatment and cleaning of the product surface prior to a painting, decorating or adhesive process is an area that has become increasingly important.[86] The processes can be aimed at simply cleaning the surfaces to be treated, or, as is increasingly common, the processes may be designed to significantly alter the adhesive characteristic and/or wettability of the surfaces.

Three of the most commonly used processes can be classified under the headings of corona, plasma and flame pre-treatments. Corona and flame pre-treatment are most commonly used for thermoplastics and will not be discussed further.

Some of the recent developments in gas plasma pre-treatments hold out more promise for the preparation of thermosetting components, since they use a high-voltage field to create a gas plasma in which dissociated particles of high energy are created, which collide with the surface of the treated component, thus breaking molecular surface bonds and creating reactive sites and thereby increasing the surface energy level of the component.[87]

9.5 Adhesive bonding

Adhesive bonding is a technique for joining materials that has increasingly shown itself capable of replacing, or supplementing, conventional approaches such as mechanical fastening, riveting or welding. Much development has occurred in adhesive formulations, means of application, and design of bonded joints to raise the status of adhesive bonding to the level of true engineering solution, which in many cases can be superior to other possible methods of joining.

9.5.1 Main advantages and disadvantages of adhesive bonding

The principal benefits of adhesive bonding include:

- the ability to join a variety of different materials, including the ability to compensate for different coefficients of expansion, elastic moduli and thickness;
- fabrication of complex shapes where other methods of fastening would not be practical;
- provide a more uniform distribution of stresses over the entire bonded area, with minimum of stress concentrations;
- elimination of irregular surface contours caused by protruding fasteners, thereby improving the surface appearance of the final product;
- the high damping capacity and elastic properties of many adhesives will allow stresses to be absorbed, distributed, or transferred, thus leading to improved fatigue resistance and vibration damping in the final product;
- potential for weight reduction — also, since no holes are required for mechanical fasteners, it is possible to maintain the integrity of structural members;
- good sealing and insulation properties, because bonded joints are sealed

against moisture and chemicals — adhesives can also act as insulators against electricity, heat or sound;
- realization of cost savings in manufacture derived from reduced material requirements, weight savings, elimination of unwanted manufacturing operations and general simplification of assembly procedures.

Limitations of adhesive bonding include:

- residual stresses in bonded joints can present problems — the stress may arise from high curing temperatures involved with some adhesives, causing problems relating to the different thermal expansion of the adhesive and the component surfaces;
- optimum bond strength is not usually realized instantaneously;
- unless toughened or flexibilized, thermosetting adhesives are likely to display brittleness with corresponding low peel strength — thermoplastic adhesives tend to creep under sustained loading;
- the long term performance of adhesive bonds in severe environments is often unknown, and so makes it difficult to assess the risks of deterioration of the joint caused by degradation of the adhesive;
- while bonded structures are not easily dismantled for repair or recovery, special tools and techniques have been developed for those situations where repair or recovery can be anticipated — motor vehicle repair for example;
- solvent-based adhesives can provide assembly operation hazards relating to flammability and/or toxicity;
- rigorous process control and testing is often needed to ensure joints of uniform quality;
- in some cases, special stages of preparation of components prior to jointing, and prolonged periods of curing under maintained heat and pressure, can prove to be onerous.

Although this list of limitations may appear substantial, developments in the technology and accumulated knowledge of adhesives and their performance is steadily minimizing many of these problems.

9.5.2 How adhesives function

9.5.2.1 Component characteristics

There are two primary features which between them will determine just how a joint based on a composite will behave, modulus and geometry, and their importance is that they dictate the stiffness of the components involved.

A component's stiffness determines to a considerable degree how stress will be distributed. This stress pattern is modified by the adhesive, and its thickness and stiffness will affect the pattern's final form. The stress distribution between the adherends and the adhesive dictates just how well a joint will perform under load and how and when it will fail when overloaded.

Composites are not normally uniform (isotropic) in nature because, apart from the reinforcing fibre, most of them have either a gel coat at the surface or, if the latter has not been deliberately created, the surface layer usually contains a higher proportion of resin than the interior. Both the gel coat and the resin-rich surface layers are brittle and are liable to display catastrophic failure when overloaded. To some degree this failing can be compensated for by an appropriate choice of adhesive. Highly compliant types are particularly good at suppressing problems of this nature because they spread the applied load over a large area and so reduce the stress borne by the friable surface of the component. Just as for composites, the relative stiffness of a metal component bonded to a composite will affect the load-bearing capacity of the joint and the way in which it will break up when overloaded. Again, but for different reasons, the nature of the surface will also influence the stability of the joint. Under bad environmental conditions (hot and humid) an adhesive/composite interface is much more stable than the equivalent adhesive/material interface. In fact, so good is the bond between the most appropriate adhesives and a clean composite that it is most unlikely that environmental stress will cause such a join to fail. This cannot be said of an adhesive/metal joint, where the adhesive and the quality of the surface preparation make a much greater contribution to long-term stability under bad conditions.

9.5.2.2 Adhesive functions

Individual atoms and molecules exert an attractive force all about them in a manner analogous to that of a magnet. They are generated by the movement of electrons and are displayed by all substances — solids, liquids and gases. It is these forces which give solids and liquids their everyday appearance and behaviour. The effect of these forces falls off very rapidly with distance. As a result, separate particles will not hold together unless they can get close enough for the forces to come into play.

Some materials, notably certain forms of rubber and gum, behave both like a liquid and a solid, the distinction between the two forms being a function of time. Because the materials are flexible and flow very slowly, these rubbers and gums will behave as separate solids when first placed one on top of the other but, given time, the two surfaces will grow together resulting in the formation of a solid mass. This type of action illustrates the action of an adhesive.

Adhesives, being liquid, flow over the surface of a solid and interact with its atomic forces as a consequence of their intimate contact. Then, as a result of their own curing processes, the adhesives turn into strong solids which, still being in intimate contact with the facing surfaces, hold them both together. The usual need for surface cleanliness is now readily apparent. Unless the adhesive is capable of dissolving or penetrating surface contamination, intimate contact will not be maintained with the underlying surface and the joint will be weakened because the required intermolecular forces cannot come into play. Similarly, if the surface is covered by a loose deposit of solid debris, intimate contact will not be obtained and again the joint will be weakened because the intermolecular forces cannot come into play properly.

No adhesive has the cohesive strength of either metal or composite, where the cohesive strength of the basic polymeric structure is reinforced by the embedded fibre and filler. So, with the exception of a gel coat, the adhesive will always tend to be the weakest link in a bonded structure. Thus, care needs to be taken to ensure that the stresses to which either it or the gel coat will be subject are well within their capacity.

While it is important to understand how adhesives function, it is equally important to understand why and how they fail. One of the primary faults has been the brittleness of the traditional, high-performance adhesives and of some of the materials used for assembling composite components. In order to achieve very high shear strength, and therefore the capacity to sustain high loads, it has been necessary to produce hard and brittle adhesive bonds. Regrettably, such materials are particularly prone to damage when the structure is shock loaded. The momentary distortion of the substrate, particularly where a flexible composite is involved, induces powerful peel and cleavage forces which these adhesives may not be able to resist.

This tendency for an overloaded joint to fail catastrophically when distorted has been substantially suppressed by the introduction of both the ductile polyurethane adhesives and the 'toughened' adhesives. The latter prevent crack propagation by incorporating a dispersed 'rubbery' phase within the harder, load-bearing portion of the adhesive. When the joint is overloaded the force which induces and propagates the crack is dissipated in the flexible rubbery phase. A similar method is used to stop a crack — that is drilling out its tip — in a sheet of over-stressed metal or plastic. Instead of the fracture energy being concentrated at the tip of the crack it is dispersed around the periphery of the drilled hole.

Although the development of the 'toughened' adhesives has been limited to various acrylic and epoxy types, the improvement in overall performance has been spectacular. Without sacrificing strength, creep or fatigue resistance, durability has been considerably increased and the peel and impact performance has been raised substantially. For these reasons the toughened adhesives are generally considered to be the prime candidates for bonding composites — especially to metal; the polyurethanes come into their own when other factors are of equal or greater importance.

9.5.3 Adhesive materials and properties

Classification of adhesive materials is no simple matter since the range of material combinations to be bonded, and the specific application performance requirements, can be endless.

Adhesive materials are available in a variety of physical forms (see Table 9.1) and for a particular application the choice is largely dictated by considerations of assembly, design, performance and cost.

Adhesive curing requirements can be conveniently classified into four main categories:

1. solvent-based adhesives, which solidify on loss of solvent;
2. temperature setting (thermoplastic) adhesives which liquefy on warming and set on cooling;
3. chemical-setting adhesives which rely on chemical reaction for curing;
4. combinations of 2 and 3.

Thermoplastic adhesives are fusible, soluble, soften with heat and are subject to creep under stress. These limitations in creep strength must be compensated for by modifying the joint design to include folded overlaps or additional stiffeners. They are, however, almost completely resistant to moisture and biodeterioration. The most important application of thermoplastics is in hot-melt adhesives, which are being increasingly employed for fast assembly, repetitive industrial operation.

Thermosetting resins solidify by polymerization through the action of heat or chemical reaction and, unlike thermoplastic materials, cannot be re-melted by further application of heat. Different chemical types have different cure requirements. In some cases, two-part resin and catalyst systems solidify by

Table 9.1 Physical characteristics of adhesives

Physical Form	Comments
Liquid	Most formulations can exist as high or low viscosity liquids, either as solvent dispersions or as aqueous emulsions and lattices In general they are easy to apply
Paste or mastic	Suitable where gap filling and non-slump characteristics are required Wide variation of consistency is obtainable
Film or tape	Restricted to use with smooth surfaces, and are characterized by ease of application and uniform glue line thickness
Powder	Usually require liquid or heat activation to arrive at liquid state Economical and have good shelf life
Granules and other solid forms	Convenient forms for ease of applications and care for particular applications Long shelf-life

chemical reaction after mixing, while other one-part materials require heat to initiate the catalyst reaction. In general, thermosetting adhesives provide stronger joints and are therefore favoured for structural assemblies. Creep properties are good, but limited peel strength and brittleness are characteristic of some types.

For the purpose of this chapter, further discussion will focus on adhesive requirements for structural applications, since outside of this field the requirements become too diffuse to allow simple treatment within the scope of this work. Those requiring more detailed information in the general field of adhesives can find such information in other specialist works.[88,89]

Where structural work is under consideration, only the toughened acrylic and epoxy types and the polyurethane-based adhesives are likely to be realistic candidates.

ACRYLICS — The two-component, toughened group

The two component versions use a variety of chemical mechanisms to initiate their 100% conversion from liquid to solid.

They are available in a variety of rheological forms and will react to give handling strengths in anything from 5 s to 30 min according to type and grade. All employ, to some degree, the use of 'rubbery' flexibilizers or the distinct concept of a secondary 'rubbery' phase within the stiff, load-bearing polymer of the major phase of the hardened adhesive. The flexible rubbery phase distorts under load and thus reduces the stress intensity at the tip of any invasive crack. This stress-reduction mechanism stops the crack from developing and thus prevents further damage to the joint.

Cure mechanism. When these adhesives were first developed one of the components could be considerd to be the adhesive proper while the other comprised the catalytic initiator of cure. They were used in a ratio of approximately 50:1 to give a 100% conversion of the liquid adhesive to solid. Because of this high ratio, and the way in which they were used without mixing, they were often called pseudo one-component adhesives. Nonetheless, the ratio required was not convenient and as a consequence there has been a progressive move towards 1:1 systems. Here, the distinction between adhesives and curing agent (the catalyst) has been lost. These adhesives do not require the precise stoichiometric proportioning of the two component epoxy-based adhesives since the influence of the one component on the other is essentially catalytic. Hence the ability to create 'dot-by-dot' and 'bead-by-bead' systems where the adhesive functions reasonably well even without proper mixing of its two parts.

As a consequence of the sophisticated manipulation of these types of formulations, curing may be initiated in any one of a number of ways; true mixing, partial mixing (dot-by-dot and bead-by-bead) and by the spreading of a catalytic initiator over one of the adherend surfaces prior to the closing of the joint. In this latter instance the adhesive proper, or the other surface, spreads over the initiator, absorbs it and polymerizes as a consequence. This film absorption has proved to be very effective in thin bond lines but only complete two-component mixing is effective when the bond lines are thick. Bead-on-bead and dot-on-dot techniques are not suitable for large gaps as there is insufficient mixing induced by the physical placement of the parts.

Normal applications. As the individual grades vary extensively in cure rate, viscosity and strength, they can, as a product group, cover a vast range of applications. At one extreme they can rival the cyanoacrylate for speed yet offer superior resistance to harsh conditions. Thus, they may be used for rapid assembly of small- and medium-sized components which need to be used in hot and possibly humid conditions. Some, particularly the toughened variants, can be used to bond load-bearing structures where gap-filling and extended cure times are essential, The designer should really give the entire product group detailed consideration before making a final selection.

Design criteria. The product group can be divided into those types which are truly toughened and the rest. The toughened group lend themselves to the assembly of substantial structures, such as bus and coach panel work, where the rather stringy viscous nature of some versions is not too much of a disadvantage. Their shear modulus lends itself to the bonding of GRP whose surfaces, especially the gel coat, is prone to be brittle. Individual cure rates and gap-filling capabilities will always tend to dictate which particular grade is most suitable for any particular application.

Adherend compatibility. Apart from rubber, and the lower surface energy plastics, most materials, and particularly the common engineering alloys, bond well. Unfortunately, zinc often presents problems unless it is either passivated or special initiators or hardening agents are used. Some types are more likely to cause stress cracking of susceptible thermoplastics than others — especially when the latter contain residual moulding stresses. In such situations prolonged contact with uncured adhesive/hardener/initiator must be avoided.

Surface preparation. Of all the adhesive types these acrylics appear to cope better with badly prepared and contaminated surfaces than any other adhesive family. Although oil contamination does reduce their performance, they can still perform well. However, this ability should be seen as a bonus which will counter poor workmanship rather than a reason for tolerating poor preparation with its attendant uncertainties. Zinc surfaces must be passivated or special initiators used and the low surface energy plastics must also be properly prepared.

Major benefits. Speed of cure and the capacity to bond to a variety of poorly prepared surfaces set these materials apart. Furthermore, some of the versions are particularly resistant to demanding outdoor environmental conditions. As a consequence, they have been successfully used to both build and repair substantial structures under less than ideal circumstances.

Major restraints. Ultimately their strength/modulus characteristics limit their load bearing capacity to a level below that of the best epoxides. Yet, in the correct design where loads are spread widely they will serve very well indeed. Unlike epoxides, their chemical nature and polymerization processes prevent the incorporation of property film modifying fillers. Hence adjustments for density, expansion, thermal transfer and electrical properties are very difficult to make; even rheological manipulation is restricted. Thus, the group as a whole is limited in this particular sense and cannot be compared with the epoxy-based materials. The latter can accommodate almost infinite subtleties of formulation manipulation.

Fault finding. Problems are usually associated with poor mixing or chemical inhibition of the polymerization processes. Where rapid cure is required some sensitivity to temperature and gap size may be noted. The liquid adhesives must be stored under cool conditions otherwise premature polymerization or other forms of deterioration may occur.

Equipment. This ranges in nature from the direct use of supplied tubes, bottles and brushes through static mixer equipped cartridges to complex fully automated systems.

Health and safety. No major toxicity problems have been associated with this family of materials and their general level of physiological activity appears to be low.

Use in accordance with the manufacturer's instructions, Health and Safety data and ensure that all COSHH requirements are in place.

EPOXIDES

This family is most conveniently reviewed by commenting on the single- or two-component versions and their toughened derivatives prior to considering general features.

EPOXIDES — The single-component, heat-cured group

These catalysed adhesives must be heated to effect cure when they form robust, thermoset and usually stiff polymers. The untoughened variants can be brittle and the more flexible versions are more prone to damage in hot, wet conditions than the toughened types. Nonetheless, the older, untoughened versions have proved popular in the past though they are being progressively replaced whenever the load-bearing, crack resisting properties of the toughened types offer advantages.

EPOXIDES — The two-component group

This group exploits the capability of the epoxide resin to react with a variety of 'hardeners' to give the very precise characteristics needed for some applications. Thus, they may be very soft and pliable when cured or very hard, brittle, cross-linked and thus resistant to severe environmental attack from aggressive chemicals. It is possible to match expansion, electrical and fire-resisting requirements very accurately. While the more pliable forms are shock resisting they are prone to environmental damage.

Most members of the two component group will polymerize at normal ambient temperature. Almost all can be warmed to increase their rate of cure though some must be heated to effect cure at all. These latter types are usually intended for use in harsh environments.

EPOXIDES — The toughened group

These adhesives owe their crack-resisting properties to the inclusion of low molecular weight, rubbery monomers in the compositions of the foregoing types. Materials such as these are more robust than the conventional single- and two-part epoxides and will resist very harsh environments. They are, as a class, one of the most robust and versatile of all the synthetic adhesives. The heat-cured versions are exceptional within the epoxy group as a whole in that they are capable of dissolving thin oil films. They have proved to be very reliable in this regard and proper industrial processes have been based upon this fact.

EPOXIDES — In general

Epoxy adhesives are thermosetting resins which solidify by polymerization and, once set, will soften but not melt on heating. Two-part resin/hardener systems solidify on mixing (sometimes accelerated by heat), while one-part materials require heat to initiate the reaction of a latent catalyst. Their properties vary with the type of curing agent — amine, amide, acid anhydride — and the resin type used. Epoxies generally have high cohesive strength, are resistant to oils and solvents and exhibit little shrinkage during curing.

Cure mechanism. Polymerization is by one of two processes. In two-component formulations, an addition reaction takes place between the active oxygen of the epoxide group and the labile hydrogen, usually from a hydroxy, amine, amide or acid group in the hardener. These reactions usually proceed at room temperature and may be speeded by warming. Some two-part versions — anhydride cured — need heating. The single-part heat-cured versions are cured catalytically and many are toughened, a process which beneficially suppresses crack propogation. All mechanisms give rise to 100% liquid to solid conversion.

Normal applications. Strength, durability and great versatility give epoxies a wide diversity of applications.

Design criteria. Because of this family's versatility it is probably more useful to summarize the limitations. A few are brittle, cure rate varies enormously with formulation, and their viscosity can make use difficult on very small assemblies. In general, their naturally very high strength may not be modified (reduced) readily.

Film-based forms are available and these can be used whenever strict weight control and accurate placement is necessary.

Adherend compatibility. This is excellent, except with thermoplastics and rubber where performance is substantially reduced. Low surface energy materials — polyolefins and fluoropolymers — may not be bonded without special surface preparation such as corona discharge.

Surface preparation. This family generally needs good surface preparation, except for a few special formulations and the heat-cured, toughened variants, which have proved very tolerant of mineral oil-based contamination. Some special versions can bond unprepared polyester composites.

Major benefits. Strength, durability, robustness and general versatility. Unlike the surface-activated acrylic adhesives, cure rate is quite independent of surface chemistry.

Major restraints. Cure rates are generally slow unless the adhesives are heated, and the special formulations which cure quickly at ambient temperatures tend to be rather brittle. The un-toughened formulations may have low peel strengths, and may exhibit brittle failure on impact. Weighing and mixing of the two-component systems must be accurate and thorough and they can be messy and difficult to clear up after use. However, special packing and dispensing systems do alleviate such difficulties.

Fault finding. Problems can arise from poor proportioning, mixing or temperature variations. Adhesive failure is almost invariably associated with contamination.

Equipment. Small quantities may be readily hand mixed and the single-part, heat-cured versions may be applied directly from a tube or cartridge. Similarly, special packing and dispensing systems may be used to dispense the two-component adhesives on the cartridge scale, but large volume metering and mixing equipment is often complex and expensive.

Health and safety. Epoxy resins and hardeners are known to cause dermatitis but, used properly, have proved to be relatively hazard-free. The dermatitic and related problems which occasionally arise with these products are often linked with either, or both, the nitrogen-based hardeners or the powerful solvents used for cleaning. Some amine-based hardeners fume strongly and must be used with effective ventilation.

Use in accordance with the manufacturer's instructions, Health and Safety data and ensure that all COSHH requirements are in place.

POLYURETHANES — Both the single- and two-component groups

The family group is named after the polymer type formed on the completion of the reaction. The adhesives are always isocyanate based. The hardeners of the two-component versions are usually based on amines or glycols. Some single-component versions can be used as hot melts. Some solvent-borne, non-reacting versions are also available.

Cure mechanism. On mixing, the isocyanate and hardener undergo a powerful, rapid addition reaction at room temperature, normally resulting in a 100% conversion to the polymeric adhesive form. However, if required, gas-producing reactions can be incorporated to foam the adhesive. This is often done to both bond and insulate. Single-component variants react with atmospheric moisture in order to effect cure. However, this reaction can take a long time to complete; 8 hours, or sometimes even more, to handling strength is not uncommon.

Normal applications. The bonding of a wide range of components fabricated from metal, wood, various plastics — both thermoplastic and thermoset — and rubber. The adhesives have general utility, but their viscosity and rate of cure can limit use on smaller components.

Design criteria. These strong resilient adhesives are capable of structural use, though the paste-like constitution of many versions prevents their ready use on small objects. The family has excellent gap-filling characteristics, particularly enhanced in foaming forms. The single-part versions are often used for both sealing and bonding when flexibility is required, and loads are either very light or nominal.

Adherend compatibility. The family as a whole is very versatile. While performance on some plastics is only modest, the group may be used with advantage on plastics materials sensitive to stress cracking. The polyolefins and other low-surface energy plastics present major difficulties unless specially prepared.

Surface preparation. May be critical and depends to a large degree on the nature of the adherend and the individual adhesive. Like cold-cured epoxies, the polyurethane adhesives do not cope well with oily surfaces. Generally, it is necessary to prime otherwise bare metal surfaces to enhance performance in hot and humid conditions. Some special versions are said to be capable of bonding unprepared polyester composites.

Major benefits. The family as a whole will bond a wide range of adherends. The various formulations may be readily manipulated to give excellent gap filling and high reaction speed coupled to a robust structural performance.

Major restraints. Both the raw materials and the final polymeric form of the adhesive are sensitive to moisture-induced damage, particularly if used on unpainted metal. Atmospheric moisture reacts readily with the isocyanate resin rapidly rendering it unusable. The cured adhesive should not be used for high loads in hot, humid environments unless surfaces have been especially prepared and the assembly has been thoroughly evaluated. The single component versions cure slowly and the rate depends on the availability of atmospheric moisture.

Fault finding. The family as a whole can suffer from inadequate surface preparation and poor wetting. Variations in reactivity can be induced by moisture, and temperature changes will affect critically the timing and efficiency of some fast-setting machine-mixed and dispensed versions.

Equipment. Modern formulations have substantially reduced the cost and complexity of the equipment required to handle these adhesives.

Health and safety. Recent developments have reduced the physiological hazards which have long been associated with these materials. Nonetheless, their use should be supported by good industrial hygiene and the very reactive versions should be treated with caution.

Use in accordance with the manufacturer's instructions, Health and Safety data and ensure that all COSHH requirements are in place.

9.5.4 Surface preparation

9.5.4.1 Composites

Were it not for the problem of mould-release agents, most new composites could be bonded without surface preparation — especially by the toughened acrylic adhesives. However, the presence of such agents must be acknowledged and taken into account. High pressure water jets will generally be sufficient for use in benign environments provided silicone release agents have not been used. If at all possible these should be avoided for they can be very persistent and difficult to remove. Light abrasion and solvent wiping is the minimum level of preparation which should be considered if bonded parts are to see robust usage and/or demanding environments. However, it should be noted that some of the newer PU formulations can bond, if heated, to unprepared composites which do contain zinc stearate mould-release agents.

Often, solvent-based primers are used in conjunction with the polyurethane-based adhesives. While these are effective and usually obviate the need for abrasion, it should be appreciated that they may well be based upon physiologically-active materials.

The durability of the adhesive/composite interface can be improved by the use of silane-based primers. These are carried in special water/alcohol mixtures and appear to be inherently safer than the isocyanate primers which they can replace. Some versions, the SIP type, offer the additional advantage that they will also indicate quite clearly whether or not the surface is sufficiently clean and well prepared to give a good and reliable result.

It should be noted also that many new analytical techniques are becoming available to assist in the characterization of surfaces for adhesive bonding.[90]

9.5.4.2 Metal

Of the three adhesive types under consideration only the acrylic variants can be used for general purpose work on unprepared metal. Even so, such a capability should be regarded as insurance rather than the licence of bad practice unless the use of oily surfaces has been very carefully assessed. This is because the toughened acrylics are capable of dissolving the residual oils and greases found on the surface of metal. The cold-cured epoxides and polyurethane adhesives are quite incapable of dissolving oil effectively. The heat-cured single-part epoxies can dissolve oil but their use on composites could probably only be justified on highly stressed components where maximum performance would be required. In such circumstances proper surface preparation should be considered essential. Clean, freshly degreased steel bonds well, and unless the metal/composite assembly is to be exposed to demanding conditions no further treatment will normally be necessary. However, if maximum performance is required then a silane primer such as SIP should be employed.

Again, freshly abraded and degreased aluminium bonds well. However, if there is to be exposure to humid conditions for prolonged periods then full surface preparation will be required. The standard chrome anodizing or phosphating processes should be employed. Consideration should also be given to silane primers since they are simpler to use than the anodizing treatments and they appear to work well on some alloys. Nonetheless, caution is advocated for not all the alloys have been assessed.

9.5.5 Design of bonded joints

The demands made on adhesive bonded joints are rarely the same as those on mechanical joints, and it should be appreciated that a simple exchange of adhesive for mechanical fastener is seldom successful. In designing the assembly it is necessary to observe two basic rules:

1. put the maximum amount of bonded area to work in contributing to joint strength; and
2. employ a favourable geometry for the joint design.

The five main types of stress encountered in adhesive bonded joints are tensile, tensile shear, torsional shear, cleavage and peel (Fig. 9.6).

Tensile loads normal to the plane of a joint can be highly destructive and must be avoided if at all possible. Indeed, every effort should be made to design a joint in such a way that it carries a compressive element within its load pattern. This is because small loads are particularly destructive when imposed through the peel or cleaving mechanism.

By comparison, compressive loads are readily borne and if these are taken into consideration then the relative ability of a joint to withstand loading in the compression, shear and peel modes is seen in the following crude, but significant, ratio — 1000 : 100 : 1. Clearly a joint should be designed to avoid peel loads. If peel and cleavage forces must be borne, then some other means of supporting the joint must be found. If this is not practical, then a complete redesign may well be required.

A common error which should be avoided is the idea that the load on a joint is uniformly distributed. Even when the most compliant adhesives are involved, this ideal state never occurs. Furthermore, the stress distribution within a joint will vary as the moduli of the adhesive and the materials being bonded vary with changes in temperature.

Mathematical models and computer programs are now available which

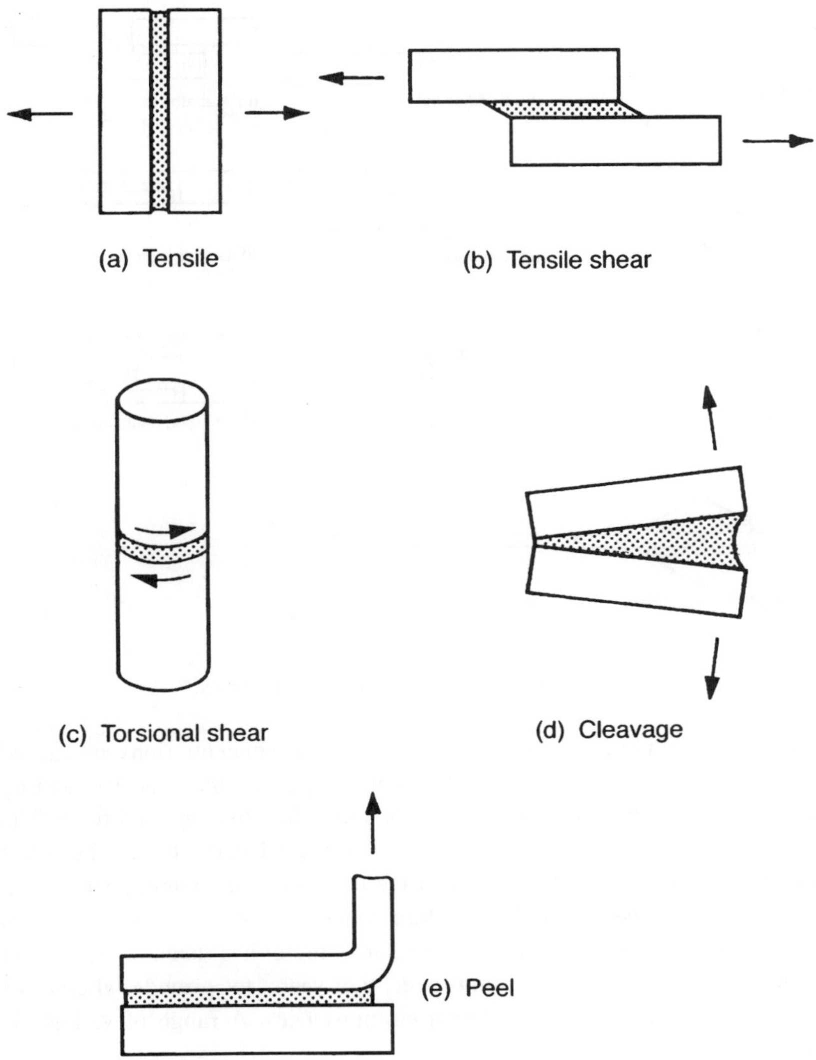

Figure 9.6 Types of stress in adhesive joints.

will enable designers to examine the interplay between the moduli of the materials, the geometry of the joint and the various dimensions involved.

Practical joints are rarely subjected to one stress, but usually to a combination of different types. Sometimes adherend distortion introduces secondary stresses. In such cases design of a favourable geometry of the joint becomes extremely important. Lap joints have been studied extensively and are the most commonly used.[88] If a simple lap joint is subjected to a tensile force then a non-uniform stress distribution will result, with stress peaks

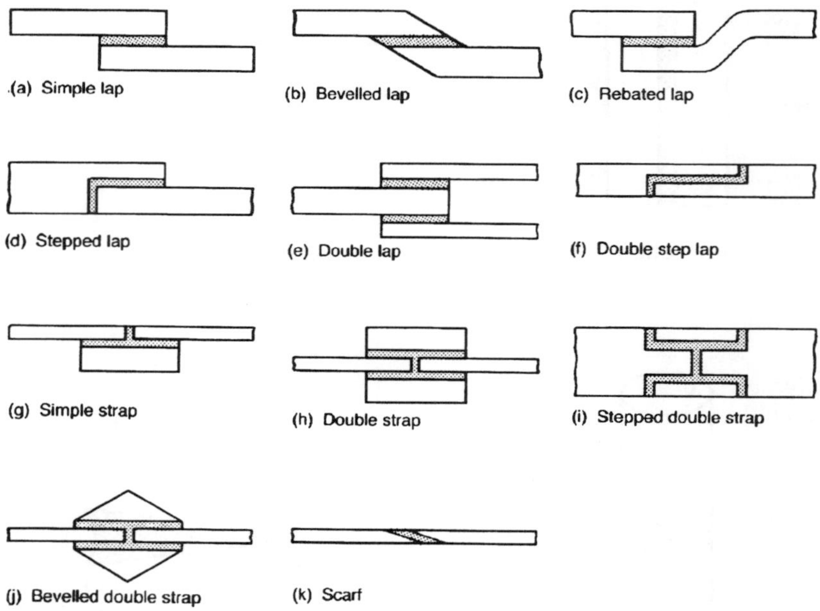

Figure 9.7 Representative types of adhesive-bonded lap joints.

occurring at both ends of the overlap. These stress concentrations are caused by the stress differential between the joint surface and the adhesive, and by the deflection of the overlapping surfaces under the offset applied force. The stress peaks will vary with the flexibility of the adhesive used. The offset loads which are typical in such a joint therefore produce bonding stresses in the bond, with risk of peeling failure where the bending movement is appreciable. For these, and other reasons, there has been considerable evolution of the basic lap joint over the years to provide alternative configurations to cope with different circumstances. A range of variants is shown in Fig. 9.7.

9.5.6 Service environments and durability

Users who are unfamiliar with adhesives frequently underestimate the damage that can be caused in hot humid conditions and often compound this error by not appreciating the types of force borne by the structure they have designed. The problem of environmental degradation is rarely associated with the bonding of the composite itself — unless it was contaminated at the time of bonding — for the thermodynamics of the situation favour the performance of the adhesive. The difficulties lie with the more polar components which may be bonded to the composite. Typical are metal, glass and ferrite. Glass is a particularly difficult surface and invariably needs either special treatment or a special adhesive if a long bond life is to be expected.

Similarly, once temperatures exceed 80°C for prolonged periods the life of many adhesives becomes quite questionable, for a variety of reasons. Again, caution is required and detailed advice should be sought. In this context, the performance of some of the toughened epoxides is noteworthy. They can often cope successfully with temperatures in excess of 150°C, even though they begin to soften at 120°C.

Shock, peel, and impact forces are often met by chance rather than by design and their effect can be quite disastrous unless the basic design concept is adequate. It is for this reason that the toughened adhesives are becoming increasingly popular. Their ability to absorb such forces is outstanding.

While the difficulties and problems should never be minimized, neither should the benefits associated with adhesive assembly be overlooked. And, in that context, the service life and performance of the toughened adhesives are particularly noteworthy. They have now been available, and in use, for enough time to demonstrate that they are capable of out-performing the designed service life of the parts they combine. It is now some 25 years since toughened adhesives were devised and used to bond the first composite trucks. Trains, cars and aircraft have followed — demonstrating quite clearly that the capability of these materials to function effectively for long periods under difficult conditions must be acknowledged.

References

1. Chanda M, Roy S *Plastics Technology Handbook*, Marcel Dekker, New York, 1986.
2. Birley A, Hawarth B, Batchelor J *Physics of Plastics — Processing, Properties & Materials Engineering*, Carl Hanser, 1992.
3. Marcantonio J *The Dynamics of the U.K. Composites Market*, 18th BPF Composites Congress Paper, The British Plastics Federation, Nov 1992.
4. Scholle K, Rowlands H *Thermoset Technologies for the Automotive Industry*, ISATA New and Alternative Materials for the Automotive Industry: Conference Paper, Florence, 1992.
5. Williamson P *Polyester Moulding Compounds, a Success Story in Underbonnet Applications*, 18th BPF Composites Congress Paper, The British Plastics Federation, Nov 1992.
6. Derek H Compression and Injection Moulding — Means of Thermoset Processing, *Kunststoffe German Plastics*, **76**: 8 (1986).
7. Ehrenstein G, Erhard G *Designing with Plastics*, Carl Hanser, 1984.
8. PLASCAMS Plastics Computer Aided Material Selector Brochure published by RAPRA Technology Ltd, Shawbury, UK, 1995.
9. West G, Tsang W *Plastics Rubber Int.* **16**: No. 3, June 1991.
10. *Guidelines to Failure Mode and Effect Analysis*, The Society of Motor Manufacturers and Traders (SMMT), London, 1989.
11. O'Neill, *Energy Balance — Automotive Materials Choices and their Impact on the Use of Natural Materials*, 17th BPF Congress Paper, The British Plastics Federation, Nov 1991.
12. Hartt G, Carey D *Economics of Recycling Thermosets*, Technical Paper, SAE International Congress and Exhibition, Detroit, USA, 1992.
13. Gentle D *End of Life Vehicle Disposal. A Car Industry Viewpoint*, Technical Paper, Plastics on the Road '92 Conference, The Plastics and Rubber Institute, 1992.
14. Kelderman H *ERCOM's Reinforced Reprocessing Starts on Schedule*, 18th BPF Composites Congress Paper, The British Plastics Federation, Nov 1992.
15. Fisher M *US Program on Automotive Recycling*, Technical Paper, Autoplas '92 Conference, 1992.
16. Saechtling H *International Plastics Handbook*, 5th Edition, Carl Hanser, Munich, 1989.
17. Billmeyer F Jr. *Textbook of Polymer Science*, 3rd Edition, Wiley/Interscience, 1984.

18. Tucker C *Computer Modelling for Polymer Processing*, 16–25, Carl Hanser, Munich, 1983.
19. Kamal M, Sourour S *Polymer Eng Sci.* **13**: 59 (1973).
20. Williamson G, Gibson A *Plastics Rubber Appl.* **4**: No. 3, 1984.
21. Castro J, Macosko C *AICLE* J. **28**: 250 (1982).
22. Cogswell F N *Polymer Melt Rheology*, George Godwin, London, 1981.
23. Kenny J, Maffezzoli A, Nicolais L *Composites Sci Technol.* **38**: 339–358 (1990).
24. Dealy *Rheometers for Molten Plastics*, Van Nostrand Reinhold, New York, 1982.
25. Brown C *Materials World*, **2**: No. 1 (1994).
26. Bagley E *J Appl. Phys.* **28**: 624 (1957).
27. Gibson A *Composites* **20**: No. 1 (1989).
28. Sourour S *Thermal and Kinetic Characteristics of Thermosetting Resins During Cure*, PhD Thesis, Dept of Chemical Engineering, McGill University, Montreal, 1978.
29. Schwartz S, Goodman S *Plastics Materials and Processes*, Van Nostrand Reinhold, New York, 15–30 (1982).
30. Modern Plastics Encyclopaedia. Published annually by McGraw-Hill.
31. Baden M Reinforcing Fibres: The Strength Behind Composites, *Materials World*, Jan 1993.
32. Klunder J *Introduction aux Composites Renforces Fibres de Verre*, PPG Fibre Glass Technical Manual. Fiber Glass bv., Holland, 1993.
33. *Thermoplastics for Toughening Phenolic Composites*, Composites Review published in *British Plastics and Rubber*, 33–35, July/Aug, 1990.
34. Katz H, Milewski J *Handbook of Fillers and Reinforcements for Plastics*, Van Nostrand Reinhold, 1978.
35. Bunsell A *Fibre Reinforcement of Composite Materials*, Elsevier Applied Science Publications, 1988.
36. Kia H (Ed.) *Sheet Moulding Compounds Science and Technology*, Chapter 4, Carl Hanser, Munich, 1993.
37. Durez Processing Data Sheets, Hooker Chemicals and Plastics Corp.
38. *News on Compression Moulding Technology*, Article in Menzolit Report, published by Menzolit GmbH, Germany, 1992.
39. Brownhill D Compression Moulding Modernizes, *Modern Plastics International*, April 1981.
40. Tellinek K, Schonthaler W, Nieman K Processing Techniques in Thermoset Moulding Compounds, *Industrial and Production Engineering*, 58, March 1979.
41. Monk J (Ed.) *Thermosetting Plastics — Practical Moulding Technology*, 83–90, George Godwin, London, 1982.
42. Gibson A *Injection Moulding of Reinforced Thermosets*, Technical Paper, PRI Conference on Thermoset Injection Moulding, Solihull, UK, 1982.
43. Gibson A Injection Moulding of Reinforced Thermosets, *Plast Rubb Proc Appl,* **3** (1983).
44. Nieman K Injection Moulding of DMC and BMC, *European Plastics News*, Oct 1987.
45. Nieman K Closed Loop Controlling the Stuffing Pressure when Injection Moulding Polyester DMC, *Kunststoffe German Plastics*, **76**: 8 (1986).

46. Johannaber F *Injection Moulding Machines. A User's Guide*, Carl Hanser, Munich, 1983.
47. Whelan A *Injection Moulding Machines*, Elsevier Applied Science Publications, 1984.
48. Choi S *Analysis of Flow, Polymerisation Rate and Moulding Cycle for Thermoset Model Compound*, Technical Paper, Society of Automotive Engineers International Congress, 1992.
49. Grant E, Leavenworth R *Statistical Quality Control*, McGraw-Hill, 1980.
50. Speight R, Yazbak E, Hull J, Coates P *Modern Sensor Technology and Benefits for Injection Moulding Process Control*, Conference Paper, NPE'94 Conference, Chicago, 1994.
51. DuBois H, Pribble W *Plastics Mould Engineering Handbook*, Van Nostrand Reinhold, 1978.
52. Austin C *Verification of Flow Analysis Software: Theory and Practice*, Technical Paper, Society of the Plastics Industry Annual Technical Conference, 1990.
53. Thomas R *The CAE Advantage for Thermosetting Materials*, Technical Paper, Society of the Plastics Industry Annual Technical Conference, 1987.
54. Indino E, Murisengo S, Molina G *A FEM Approach of Thermoset Materials Injection Moulding Modelling*, FPGP'94, 475–493, 1994.
55. Beaumont J, Koch P, Farrell R *Frictional Heating in Runners for Enhanced Thermoset Moulding*, Technical Paper, Society of the Plastics Industry Annual Technical Conference, 1990.
56. Hain D, McCaffrey N *Using Computer Aided Engineering to Analyse Runner Designs for the Integrated Circuit Industry*, Paper produced by Mold Flow Pty Ltd, Kilsyth, Australia.
57. Beamont J *Evaluation of Runner Design Through Use of Thermoset Flow Analysis*, Technal Paper published by Moldflow Inc. Kalamazoo, Mich. USA.
58. Burns R *Polyester Moulding Compounds*, Marcel Dekker, New York, 1982.
59. *Composites for the Future*, Congress Papers of the 18th International BPF Composites Congress, Pubn No. 293/6, The British Plastics Federation, 1992.
60. Signorelli E *Research and Development in Injection Moulding BMC*, Technical Paper published by Savid S.p.A, Italy1986.
61. Demez K *Injection Moulded BMC Automotive Parts: Higher Performance Through a Total System Approach*, Technical Paper, Society of the Plastics Industry Inc, ANTEC Conference, 1991.
62. Demez K *The Outcome of a Several Year's Program: The ZX Hatchback*, Technical Report published by Owens Corning Fiberglass, Belgium, 1992.
63. Michaeli W, Heber M *Calculation of Fibre Orientation for Complex SMC Parts*, Technical Paper, Society of Plastics Engineers ANTEC Conference, 1994.
64. Kia H *Sheet Moulding Compounds Science and Technology*, Carl Hanser, Munich, 1993.
65. *SMC Design Manual — Exterior Body Panels*, published by The Composite Institute of the Society of the Plastics Industry Inc, 1991.
66. Coissac J High Pressure In-Mold–Coating Paints, *Composites*, No 2, March/April 1992.
67. Becker W (Ed.) *Reaction Injection Moulding*, Van Nostrand Reinhold, New York, 1979.

68. Coates P, Johnson A An Introduction to Reinforced Reaction Injection Moulding, *Plast Rubb Process Appl* **1**: 223–228 (1981).
69. Oertel G (Ed.) *Polyurethane Handbook*, Carl Hanser, Munich, 1985.
70. Manas–Zloczower I, Macosko C *Poly Process Eng*, **4**: 173–184 (1987).
71. Sweeney S *Reaction Injection Moulding Machinery and Processes*, Marcel Dekker, New York, 1987.
72. Woods G *The ICI Polyurethanes Handbook*, John Wiley & Sons, New York, 1987.
73. Macosko C *Fundamentals of RIM*, Carl Hanser, New York, 1989.
74. Klempner D, Frisch K *Polymeric Foams*, Carl Hanser, Munich, 1991.
75. Tucker C, Suh N *Polym Eng Sci* **20**: 875–886 (1980).
76. Lee L, Ottino J, Ranz W, Macosko C *Polym Eng Sci* **20**: 868–874 (1980).
77. Kolodziej P, Macosko C, Ranz W *Polym Eng Sci* **22**: 388–392 (1982).
78. Sandell D, Macosko C, Ranz W *Polym Process Eng* **3**: 57–70 (1985).
79. Coates P, Sivakumar A, Johnson A *Plast Rubb Process Appl* **8**: 203–213 (1988).
80. Coates P *Plast Rubb Process Appl* **13**: 139–146 (1990).
81. Muller M, Mrotzek W, Menges G in *Reaction Injection Moulding*, Ed. J. Kresta, *ACS Symp Ser,* **270**: 237 (1985).
82. Maier U, Menges G *Soc Plast Eng Technical Papers* **44**: 1350–1355 (1986).
83. Coates P, Johnson A *Reinforced Reaction Injection Moulding: Exploring the Machine–Materials Interface*, Proc. UTech '86, 49–54, Crain Communications Ltd, London, 1986.
84. Coates P, Johnson A, Armitage P, Hynds J, Leadbitter J *Polym Eng Sci* **27**: 1209–1215 (1987).
85. Padilla J *Abrasive Water Jetting*, The Welding Institute Research Bulletin, Vol 26, Oct 1985.
86. Sullivan G *Pre-treatment of SMC for Painting*, Technical Paper, Society of Automotive Engineers Congress, Detroit, Feb 1992.
87. Kaplan S, Rose P *Plasma Surface Treatment of Plastics to Enhance Adhesion — An Overview*, Technical Note published by Plasma Science Inc, Belmont, California, 1990.
88. Shields J *Adhesive Bonding*, Published by The Design Council/British Standards Institute, London.
89. Lees W *Adhesives in Engineering Design*, published by The Design Council, London.
90. Madura A *Multi-technique Surface Characterisation of SMC for Adhesive Bonding Applications*, Technical Paper, Society of Automotive Engineers International Congress, Detroit, Feb 1992.

Index